Can Crafts

Can Crafts

EARL FOWLER

Drawings by Joan Holmes

CHILTON BOOK COMPANY • Radnor, Pennsylvania

Library of Congress Cataloging in Publication Data

Fowler, Earl P.
    Can crafts.

    (Chilton's creative crafts series)
    1. Tinsmithing.   2. Tin cans.   I.   Title.
TT266.F67   1977   745.56   77-3522
ISBN 0-8019-6233-1
ISBN 0-8019-6234-X pbk.

1 2 3 4 5 6 7 8 9 0   6 5 4 3 2 1 0 9 8 7

Contents

Contents

Preface

Ecology—recycling waste material—and the subject of what to do with leisure time are in the news quite a bit these days. I am not foolish enough to believe that this book will solve all these problems, but I surely hope it will inspire the reader to bend over, pick up an old tin can, and recycle it into a beautiful and useful article.

Tin can crafting is an interesting, creative and, most of all, inexpensive hobby. Some people have the mistaken idea that elaborate tools and expensive equipment are needed to create beautiful tin can crafts. Not so. This book was written with the intention of showing how to keep expenses down. The instructions are simple and the illustrations plentiful. It is hoped that this will appeal to the busy young homemaker, as well as to the retired person living on a fixed income.

Some people shy away from crafting with cans for fear of cutting themselves. I have been working with tin cans for five years and I have yet to get a bad cut—a few scratches maybe, but no bad cuts. The serrated blades of the shears I use leave a dull edge on the tin work. Do be careful, though, not to leave little splinters along the edges of your cut work. A prick with a splinter of tin can be as sore as a bad cut.

People ask, "How many times do you cut your fingers?" My reply is a question: "How many times do you burn your fingers when you strike a match?" Respect the cans, watch how you handle them, and I am sure you won't need Band-Aids in your toolbox. If you think you will be safer wearing gloves, by all means do so.

My wife and I retired in 1965 and started traveling around the country. As we toured the Great Northwest, we visited a state fair where I met and fell in love with "beautiful tin cans."

My hobby has grown and my diet has changed a lot since 1965. You see, I do the grocery shopping now. I don't look for bargains in the beat-up basket, nor do I look at the labels on the cans. I search for and pick up the can that can best be used for the project I have in mind at the time. My wife has changed many an evening menu—not for nutritional reasons, but just to empty a can for me.

When *your* shopping and eating habits start to change, look out: The "can-bending bug" has got you!

The projects in this book were not just selected at random. They were chosen to show the different techniques and types of cans used to make can crafts. If some of these projects don't strike your fancy, please don't just flip the page and go on. Stop a minute and study the plans; they may give you ideas and inspiration for a project of your own.

ix

Acknowledgments

To my wife, Willie, for putting up with me for the last forty years (in general) and the last eighteen months (in particular). She deserves a citation for her ability to make delicious meals from the contents of odd-shaped cans.

To my daughter, Susan C. Kallmeyer, for her ability to decipher my notes and handwriting while typing and retyping the manuscript.

To my son, James E. Fowler, for his photography and proofreading.

To my teenage granddaughter, Denise, for her creation of the milk can candle holder.

To our dear friend and neighbor, Joan E. Holmes, who devoted endless hours creating the many tedious drawings.

And last, but not least, to our good friend and adviser, Hazel Fuller of Widbey Island, Washington, for sharing her designs and teaching me several of her techniques.

Can Crafts

Chapter 1

Basic Tools and Skills

All the instructions and projects in this book have been tried and proved. They all work, but this does not mean that you have to follow my directions to the letter. Experiment with different tools and try different methods: You may come up with a way that is more convenient and comfortable for you.

TOOLS AND EQUIPMENT

The tools required for this craft are few and, for the most part, inexpensive. Most can probably be found in your workshop right now. The most important tool is a *good* pair of tin snips or shears which will cost $5 or $6—well worth it when you think of the hours of trouble-free enjoyment they will bring. Here are the tools you will need most frequently (Fig. 1–1).

Lobster Shears—These shears are easier to use than regular tin snips. They can be purchased at most hardware stores. If not in stock, have them ordered for you (Stock WISS No. C4-ARL).

Tin Snips—The 7″ size is best for cutting tough seams and rims. Snips may be used in place of lobster shears.

Can Opener—A wall-type opener or a good hand model is a necessity.

Awl or Icepick—For marking and punching holes.

Needle-Nose Pliers—A pair with a 2″ or 3″ nose is fine; a pair with side cutters will come in handy.

Ruler and/or Tape Measure—A 6″ or 8″ ruler for straightedge and measuring.

Ball-Peen Hammer—Small, for shaping some flower petals.

Compass and/or Dividers—For drawing circles.

Small Magnet—Not a necessity, but comes in handy for picking up small bits of tin that might prick the fingers if you tried to pick them up by hand. Keep the magnet out of your toolbox, as it might magne-

1

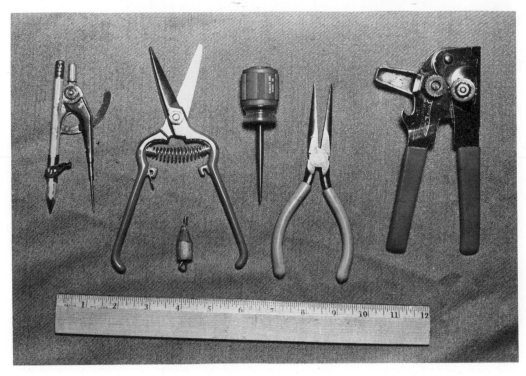

Fig. 1–1 Tools used most frequently: *from left*, compass, tin snips, awl, needle-nose pliers, can opener; *below*, homemade curling tool and ruler

tize the ends of your pliers and snips. This can be bothersome at times.

Curling Tools—These can be made at home. Two methods for making your own are discussed later in this chapter under Curling.

Some of the other supplies that you may need include:

Band-Aids—Just in case!

Decorative Beads

Floral Tape—Available at dime stores and craft and hobby shops.

Gloves—If you feel you need them.

Nuts and bolts—6 x 32 x ½" size is best.

Paint—See Painting section later in this chapter.

White glue—Such as Elmer's.

Wire—Coathangers, 28-gauge floral wire, 16-gauge wire are each used for different purposes.

Did you notice that a soldering iron isn't listed? Why solder when you can get by without it? However, if you have one and know how to use it, it may come in handy on a few of the projects.

TYPES OF CANS

The grade or quality of steel varies from can to can. Once you start to work, you'll soon learn which soup, vegetable, fruit, or beverage can you prefer.

Some projects in this book call for cans with ridges. (In the canning industry these ridges are called beads; their purpose is to make the can more rigid.) In most cases, though, it is better to select cans with smooth, straight sides. They are easier to mark and cut; if there is curling to be done, the end product is much neater (Fig. 1–2).

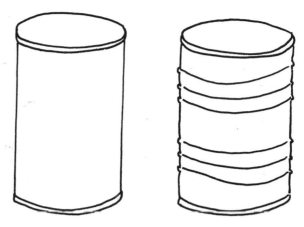

Fig. 1–2 Smooth can and can with ridges or beads

The steel used in soda and beer cans is brittle and breaks away from the base too easily. Avoid these cans in projects that call for cutting strips down to the base, or for making sharp bends in the tin.

I use aluminum cans only for some flowers. It's easy to tell the difference, as steel cans have a side seam not found on aluminum cans.

PAINTING

It is recommended that all finished projects be painted or sprayed with a protective coat of lacquer or clear plastic. Tin cans are sheets of steel, plated with a thin coat of tin. Cutting a can breaks through the tin plate and exposes the steel. If not protected, exposed areas will soon rust, ruining a beautiful object of art representing many hours of work.

Generally, I use oil-based spray paint and prefer Accent matte craft paints, which leave a soft, flat finish. For black, I use "wrought iron black." To keep costs down, buy paints on sale or at discount department stores.

Most cans have a gold-colored lining of lacquer or resins used to protect the food contents from a chemical reaction with the tin. No two food products have the same reaction to tin, so different types of resins are used for liners. Some of these resins may blister when painted with enamels. If so, remove the paint and rub the resin off with fine steel wool, then repaint. You will soon learn which cans give you trouble: Avoid them.

I use aluminum cans to make flowers and leaves for several reasons: Aluminum cans are soft and can be cut with ordinary scissors, which means more flexibility when cutting intricate shapes. Flower petals and leaves look more natural when shaped with aluminum. I like to paint my flowers with acrylics so I can mix my own colors and shades. Acrylics are water-based paints and might cause rust if used on steel. After painting with acrylics, brushes are easily cleaned with soap and water.

WORKING WITH CANS

When preparing a can for use, peel off the paper label. Don't be concerned if the label is printed directly on the can: When you paint the finished project, it will be covered. If there are spots of glue left on the can, scrape them off while dry instead of trying to wash them off. (Fingernail polish remover will take off some types of glue.)

No matter what size the can, the top rim and the side seam must be removed. To do this, lay can on a flat surface and place the rim in the cutting edge of the can opener. As your can opener is being turned, make sure to keep pressure on the can with forefinger (Fig. 1–3a). If this is not done, the can will have a tendency to turn, causing cutter to leave the rim (Fig. 1–3b).

Trimming and Removing Seams

If the can opener has left a rough edge around top of can, now is the time to trim it (Fig. 1–4). Be careful at this point: The can opener does not leave a dull edge, as do the lobster shears. *Don't* use your fingers to straighten the edge; *use pliers*.

You will have better control and more leverage if, when cutting, you make short strokes using the center section of the blade (Fig. 1–5a). Try not to run the shears clear down to their points, except when you reach the end of the strips at the base of the can (Fig. 1–5b). Cut as close as you can to the base (go slowly) without letting the tips of blades snap. If you allow the points to meet, the tin has a tendency to tear (Fig. 1–5c).

To cut the seam out, place your shears as close as possible to the right side of seam and cut to base of can (Fig. 1–6a). Now cut down the left side. To remove the seam, grip it just as close to the base as you can with your pliers (no fingers, please) and bend it back and forth (Fig. 1–6b).

Dividing Sections and Strips

Most projects call for the can to be divided into equal sections or spokes. An easy way to make these divisions is to cut a strip of paper (brown grocery bags are good). Wrap this strip of paper around the can, starting at one edge where seam was removed and ending at the other edge. Mark and cut strip at this point (Fig. 1–7a).

Now fold the strip in half; fold in half again for four sections; in half again for eight sections and, if sixteen sections are needed, fold in half again. Mark at the folds (Fig. 1–7b).

Replace the paper around the can and mark. Now move the paper to the base of the can and mark again (Fig. 1–8a). Use an awl or icepick to scratch lines (Fig. 1–8b). If you have trouble seeing these lines, try using a grease pencil, such as grocers use.

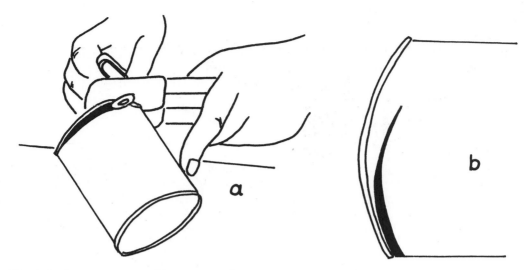

Fig. 1–3   Removing top and bottom rims from a can

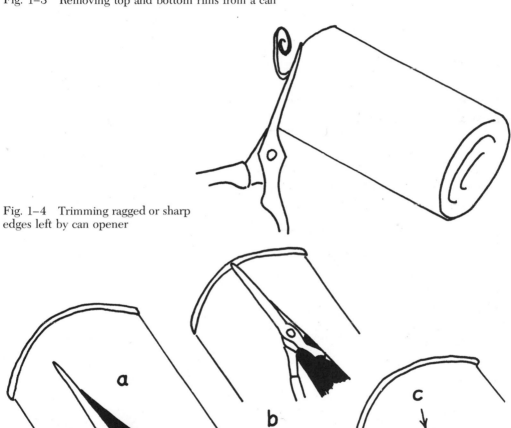

Fig. 1–4   Trimming ragged or sharp
edges left by can opener

Fig. 1–5   Techniques for cutting cans

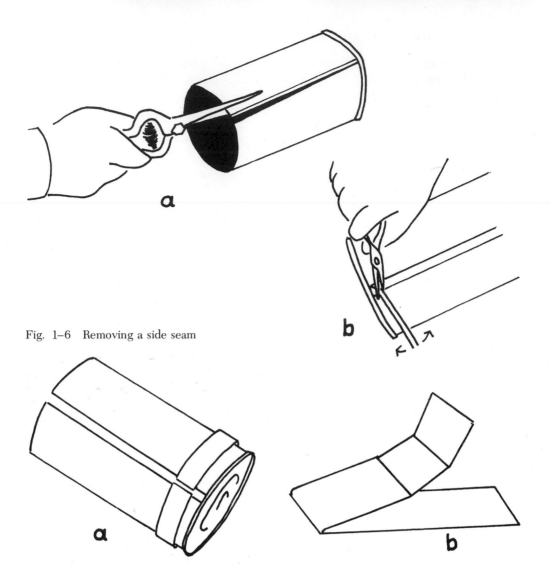

Fig. 1–6　Removing a side seam

Fig. 1–7　Dividing a can into equal-sized sections

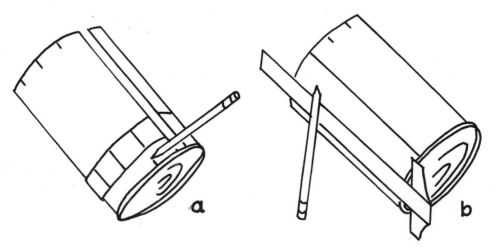

Fig. 1–8　Marking a can for cutting into equal sections

After the can has been cut into the desired number of sections—usually eight, twelve, or sixteen—you will have to divide them into narrow strips. If your project calls for eight strips per section, just divide the section into halves, fourths, then eighths (Fig. 1–9).

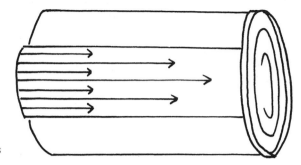

Fig. 1–9   Dividing sections into strips

Bending and Shaping

Now that you have the can cut, you are ready to start bending and shaping it. To position the strips for working, grip each one with pliers about 1/4″ from base and give one-quarter turn, either to clockwise or counterclockwise, whichever is called for in the project you're working on (Fig. 1–10a). If you grip the strip too close to the base of the can when turning, you may tear it loose. Do be careful (Fig. 1–10b).

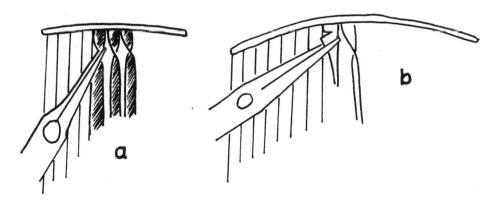

Fig. 1–10   Using needle-nose pliers to turn strips

When directions call for grouping or gathering several strips together, simply pull the prescribed number together with your pliers and clamp (Fig. 1–11).

Making Clamps

To make clamps, cut strips the same width as the strips you are clamping and about 3/4″ long. (For example, the strips in the heart basket are about 1/16″, so the clamps should be the same width. The strips used for the trivets are 1/4″ wide, so clamps the same width should be used. I think they look better.) Bend a

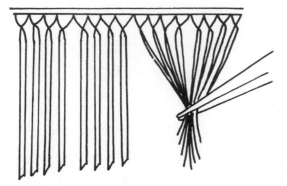

Fig. 1–11   Gathering strips into groups

U in the end of each clamp, using your needle-nose pliers (Fig. 1–12a). Place the clamp around the group of strips that is to be gathered together (Fig. 1–12b). Draw tight, squeeze with pliers, and cut off excess (Fig. 1–12c).

Clamps or ties should be placed at uniform height. Rather than repeatedly measure with a ruler, why not make yourself a gauge based on the height of the first clamp in a series? Make a gauge out of scrap tin or cardboard (Fig. 1–13).

Curling

Curling is achieved by putting the end of a strip into the slot at the end of a curling tool (Fig. 1–15a). Here are two ways to make your own.

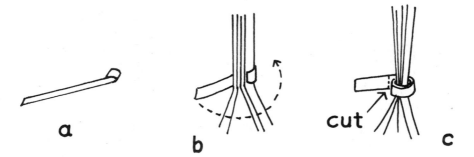

Fig. 1–12   Clamping strips together

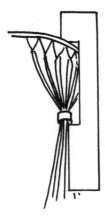

Fig. 1–13   Using a gauge or template for marking clamp position

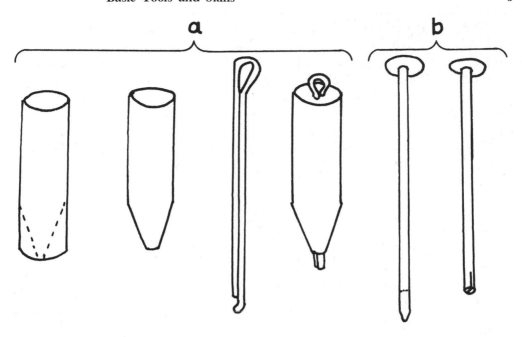

Fig. 1–14   How to make curling tools: *at left*, dowel sticks with cotter keys; *at right*, nail with slit in the end

1. Drill small holes in short lengths of dowel stick, $1^1/_2''$ to $2''$ long. Push cotter keys (pins) through the holes (Fig. 1–14a). Dowels in different diameters should be used: You'll need $^3/_4''$, $^5/_8''$, and $^1/_2''$.

2. Remove the point from a large nail and cut a slit in the end with a fine hacksaw blade (Fig. 14b).

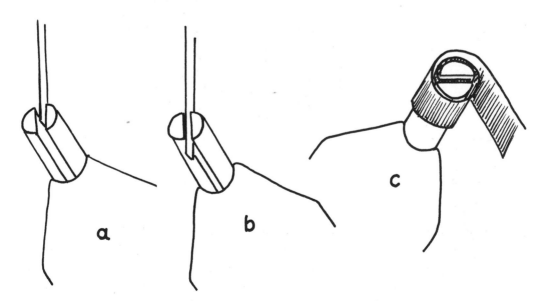

Fig. 1–15   Curling tin strips with a curling tool

With curling tool in hand, put end of strip into slot. Do not let the strip protrude beyond the lip of the tool (Fig. 1–14b). If it does, the curl will be egg-shaped. Hold the strip and the tool firmly, keeping the tool at right angles to the strip. This enables the strip to be curled neatly on top of itself (Fig. 1–14c).

An easy way to divide a lid or disk into equal sections is to cut a circle of paper just a bit smaller than the lid you intend to use (Fig. 1–16a). Fold the paper in half once (Fig. 1–16b), twice (Fig. 1–16c), and then three times (Fig. 1–16d). Mark the folds (Fig. 1–16e), lay the paper back on the metal disk, and mark.

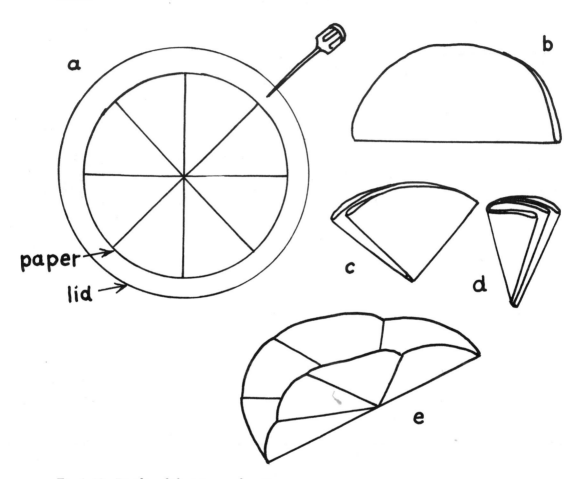

Fig. 1–16   Dividing disks into equal sections

Joining Cans

There are several ways to join two cans together. Probably the easiest way is to bolt them together (Fig. 1–17a). If you do this, make certain the holes for the bolt are centered so the cans will make a smooth joint. The second way is to cut a disk of corrugated cardboard 1/8″ smaller than the base of the can, put Elmer's glue on both sides of cardboard, and slip between the can bases (Fig. 1–17b). The third method is to solder.

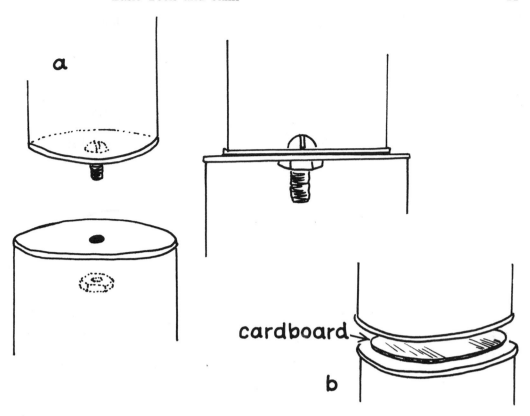

Fig. 1–17   Two methods for joining cans together

If you would like to dress up the joint where the two cans unite, why not glue a piece of rickrack or braid of some sort over it (Fig. 1–18). This may be added either before or after painting, depending on the decor you desire.

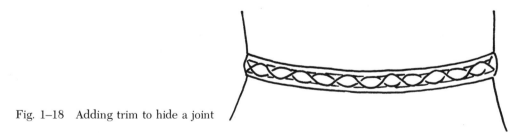

Fig. 1–18   Adding trim to hide a joint

Handy Hints

When preparing an aluminum can for use, the bottom *cannot* be removed with a can opener. Cut a slit down the side of the can to within $1/4''$ or $1/2''$ of the bottom. Turn your blades parallel to the base and cut off (Fig. 1–19a). Trim up the edge around the bottom and save it for future use. Hold the sheet of aluminum in both hands and draw over edge of table (Fig. 1–19b). This tends to take the curl out and makes it easier to work with.

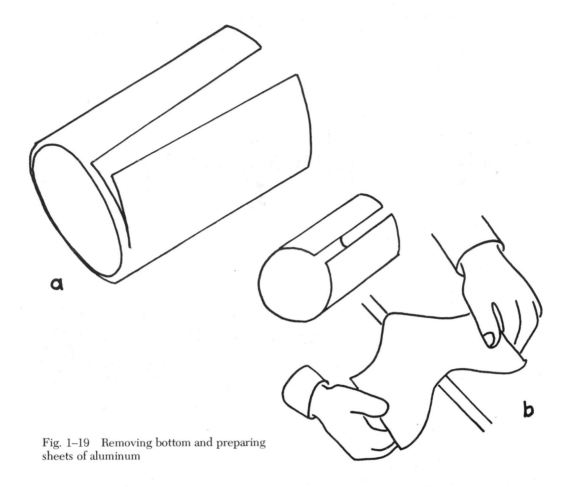

Fig. 1–19   Removing bottom and preparing
sheets of aluminum

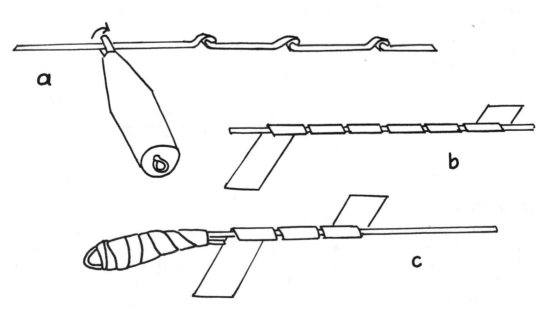

Fig. 1–20   Making fancy trims

12

The fancy trims for top rims of baskets? Here are two simple methods.

1. Cut strips from a 1-gallon oil can. Cut these the width of the strips they are to be attached to. Mark the strips at regular intervals (1″ or 1¼″). Place curling tool on line and give one full turn (Fig. 1–20a). Bind to basket with floral wire.

2. Cut strips ³/₁₆″ wide. Spiral them around a piece of coathanger wire (Fig. 1–20b). A loop on one end will give you a good grip (Fig. 1–20c).

Chapter 2
Beginner's Corner

Now that you have read and familiarized yourself with the general directions, I assume you are ready and anxious to get started. If you are a beginner and this is your first time at cutting tin, it would be helpful to gather up some cans and practice cutting and curling just to get the feel of it.

The projects in this chapter are shown in Fig. 2–1; they are all very easy and simple. Even if some do not appeal to you, make them up for practice and to learn the different techniques.

Fig. 2–1   Beginner's corner projects: *from left*, holiday bell, milk can candleholder, Christmas wreath, sconce, pincushion or footstool, and ashtray

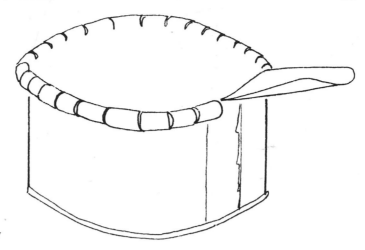

Fig. 2–2   Ashtray

ASHTRAY

For this project, shown in Fig. 2–2, you will need one beverage can (*not* aluminum).

Remove top and top rim, but do not remove side seams. Shorten can to 2³/4″ and put a mark around can 1″ from bottom (Fig. 2–3). Draw a line ¹/₂″ to each side of the seam and cut down to the 1″ line (Fig. 2–4). Using these two cuts as a starting point, divide the rest of the can into ¼″-wide strips (Fig. 2–5). The total number of strips is not important. Cut these strips down only to the 1″ mark.

Curl all these narrow strips down, using needle-nose pliers or a large curling tool (Fig. 2–6).

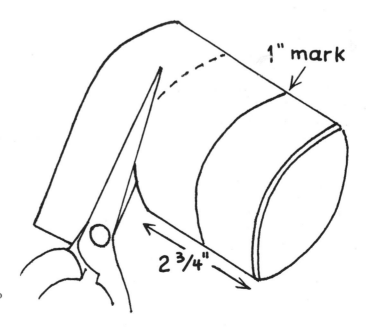

1″ mark

2³/4″

Fig. 2–3   Cutting ashtray to 2³/4″ height

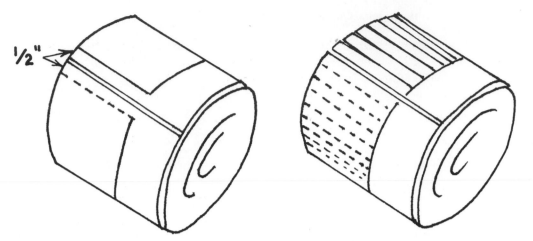

Fig. 2–4 Can measured and ready to have strips cut

Fig. 2–5 Cutting ¼″-wide strips

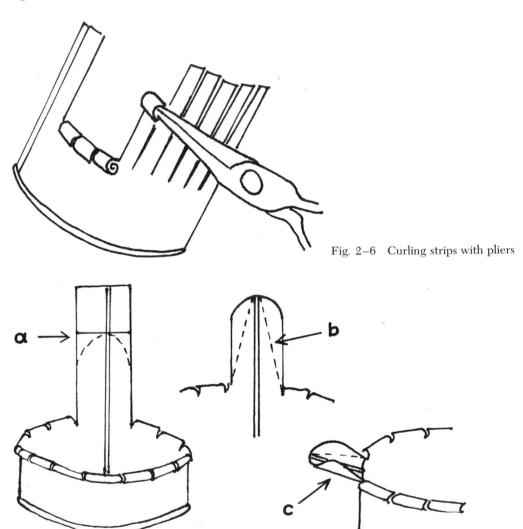

Fig. 2–6 Curling strips with pliers

Fig. 2–7 Putting finishing touches on ashtray

Cut off about 1/2″ from the section with the seam, rounding the corners as shown in Fig. 2–7a. With needle-nose pliers, bend the sides up (Fig. 2–7b). Now, using your fingers, bend this section down 90° (Fig. 2–7c).

SCONCE

For this project (Fig. 2–8), you will need one soda or beer can (*not* aluminum—it's too soft).

Remove top and top rim. Draw two lines around can: one 1″ from the bottom, and one 3″ from the bottom.

Measure and mark 1¼″ from each side of the seam. Draw and cut a line from top of can down to the 1″ line (Fig. 2–9). Thus the seam will end up in the center of the finished handle.

You will now have two sections: The one with the seam is 2½″ wide; the other is 5¾″ wide. Cut 1½″ off the top of the larger section. Now divide this section into strips approximately 5/16″ wide. Total number of strips is not important.

Use a large curling tool and curl all these narrow strips down toward you (Fig. 2–10).

Go back to the 2½″-wide section with the center seam. Cut five 1/8″-wide strips down on each side. Terminate each cut of the first strip 1/2″ from the 1″ line

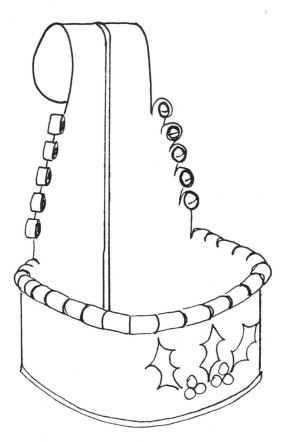

Fig. 2–8 Sconce

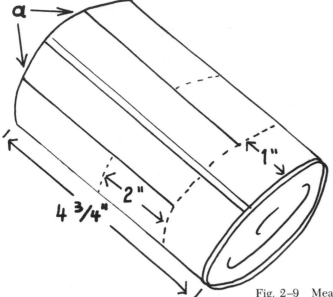

Fig. 2–9   Measured and marked can, ready to cut

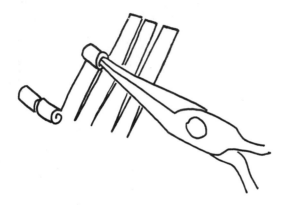

Fig. 2–10   Curling down strips for sconce

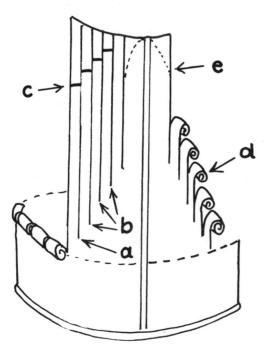

Fig. 2–11   Back of sconce, showing depth of successive cuts

Fig. 2–12   Handle and side of finished sconce

(Fig. 2–11a). Terminate each of the succeeding cuts ¹/₄″ shorter than the previous cut (Fig. 2–11b).

Make each strip 1¹/₂″ long from base of cut to tip (Fig. 2–11c). Curl them down and turn them out. (Fig. 2–11d).

Round off the corners of the remaining strip (Fig. 2–11e). Bend it back in a large arc to serve as a handle (Fig. 2–12). Paint for a decorative finish.

MILK CAN CANDLEHOLDER

To make the candleholder shown in Fig. 2–13, you will need one beverage can (*not* aluminum), and 15 inches of fine wire.

Remove top and top rim of can. Draw a line around the can 2″ up from the bottom (Fig. 2–14a). Measure and mark ¹/₂″ from each side of the seam. Draw and cut a line from the top of can down to the 2″ line (Fig. 2–14b). This 1″ strip, with the seam in the middle, will be the handle. With your needle-nose pliers, curl this strip down into a U turn (*see* Fig. 2–13a).

Cut strips ¹/₄″ wide around rest of can, cutting down to the 2″ mark (Fig. 2–14c). This should give you about twenty-eight strips (total number of strips is not important). Number each strip, as shown in Fig. 2–13.

Place a large curling tool at the end of each *even-numbered* strip and curl one and one-quarter turns toward you. Thread these curls onto a piece of fine

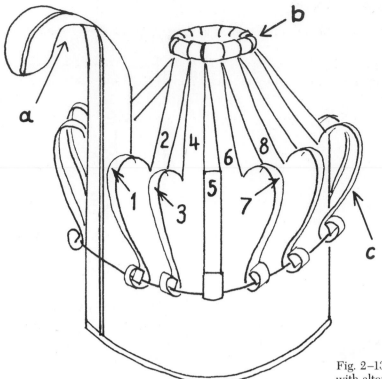

Fig. 2–13   Milk can candleholder with alternating strips

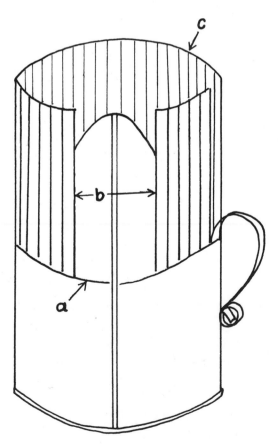

Fig. 2–14   Can cut into strips, ready to curl

wire and draw them into the center of can in a circle (Fig. 2–13b). Twist wire to hold.

Place the curling tool at the end of each of the *odd-numbered* strips and curl one and one-quarter turns away from you. Bend these strips down toward outside of can in a large graceful arc. *Do not* bend in flat against can (Fig. 2–13c). Thread these curls onto a piece of fine wire and draw *only* the curls in tight against the can.

Paint if desired. Add a tall candle.

CHRISTMAS WREATH

For this project, shown in Fig. 2–15, you will need one $2^5/_8''$-diameter soup can; one small angel or Christmas ball; small colored beads.

Remove top and top rim from a $2^5/_8''$ can. Remove bottom. Remove side seam. Shorten can to about 2" from bottom.

Divide can into eight equal sections (*see* Ch. 1). Divide each section in half. Divide each of these sixteen halves into five equal strips. Lay can on a flat surface and bend all eighty strips out flat to a 90° angle.

Start at the space where the seam was removed and work to the right. Twist two strips counterclockwise; then twist three strips clockwise. Clamp these five together, $^1/_2''$ from bottom of can (Fig. 2–15a). Curl three strips down to the right and two strips down to the left (Fig. 2–15b).

Fig. 2–15 Christmas wreath

Work all the way around the can, twisting, clamping, and curling in groups of five.

Paint; attach a string of colorful beads (the color you desire) around the outside rim of can. Hang a small tree ball, angel, or some other Christmas ornament in center.

FOOTSTOOL OR PINCUSHION

Materials needed for this project are: one $2^1/_8$"-diameter juice can; one 2" disk, a piece of padding, and a piece of velvet for a cushion.

Remove top, top rim, and side seam from the can. Shorten can to about 3" from bottom. Divide can into eight equal sections, then divide each section into six equal strips (*see* Ch. 1).

Starting at space where seam was removed, hold can bottom-side-up and, working toward the right, twist three strips one-quarter turn clockwise and three strips one-quarter turn counterclockwise (Fig. 2–16a). Gather these six strips; curl them in good and tight to make a stable leg, as shown in Fig. 2–17a.

Moving to the right, skip the next six strips. Come back to them after all four legs have been completed. Take the third group of six and make another leg in the same manner as you made the first. Skip six strips. Take the fifth group of six strips and make another leg. Skip six, then make the fourth leg from the next group.

After the four legs have been completed, you should have four groups of six strips each remaining. Shorten these strips to about $1^1/_2$" from bottom of can. Bend every other strip straight up, then twist all strips one-quarter turn clockwise (Fig. 2–17b). Curl three up and three down, tight against the rim into the vacant spots (Fig. 2–17c). Paint.

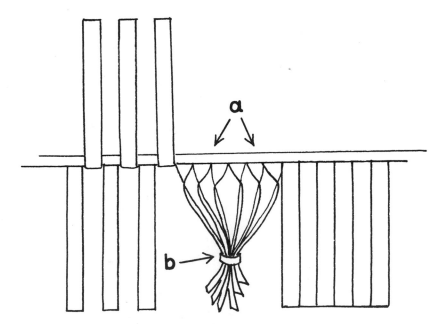

Fig. 2–16   Footstool or pincushion with strips cut and ready to shape

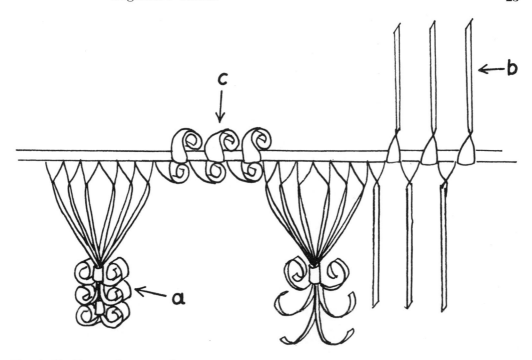

Fig. 2–17   Footstool or pincushion with strips curled in groups of six

Cushion: Cut a tin or cardboard disk about $1/8''$ smaller in diameter than the bottom of stool. Cut a piece of foam rubber, sponge, or Styrofoam for cushioning. Cover with a piece of scrap material (Fig. 2–18). Glue to stool.

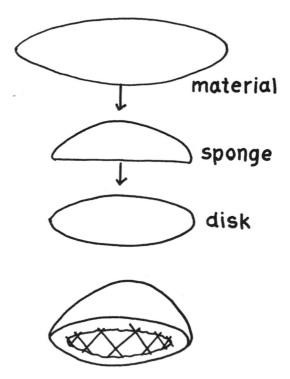

material

sponge

disk

Fig. 2–18   Assembling the cushion for footstool

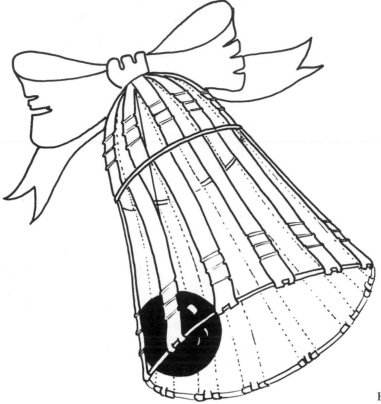

Fig. 2–19   Holiday bells

HOLIDAY BELLS

To complete this project, shown in Fig. 2–19 and in color Fig. 12, you
will need: one $3^1/_8''$-diameter can about $4^1/_8''$ tall (use a can with
plenty of ridges); one rim from a 2-pound coffee can (or 3-pound
shortening can); one Christmas tree ball, $1^1/_2''$ or smaller; nut and
bolt; wire, red ribbon, and paint.

Remove top and top rim of $3^1/_8''$-diameter can. Remove bottom. Draw a line
around can $1/_2''$ from the top. Remove side seam. Divide can into eight equal sections. Divide each section into three equal strips (see Ch. 1). This will give you
twenty-four strips, each about $3/_8''$ wide.

Hold the can bottom-side-up; number the strips 1 through 24, starting at the
space where seam was removed and working from left to right. Bend all the
even-numbered strips upward. Bend them in behind rim and up through center
of can (Fig. 2–20). Shorten these even-numbered strips to about $3^1/_2''$. Round the
ends off (Fig. 2–20a). Punch a small hole in the end of each strip. Bolt or wire
these ends all together, forming them into a dome as you bolt them (Fig. 2–21).

Cut two slits in the end of each odd-numbered strip to divide the end into
thirds. Cut these slits to the $1/_2''$ line drawn around the can earlier (Fig. 2–20b).

Cut the rim off a 2-pound coffee can (or a can of similar size). After rim has

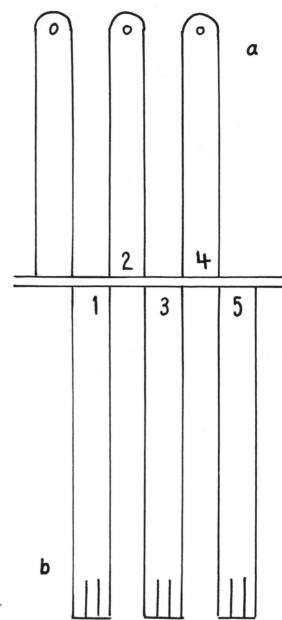

Fig. 2–20   Strips cut and bent to position,
ready to shape

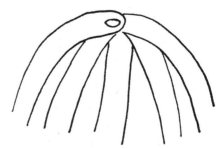

Fig. 2–21   Ends of even-numbered strips gathered
into a dome

25

been removed, go all the way around with a pair of pliers, crimping or pinching down any sharp edges. This rim will form the bottom lip of the finished bell.

At the end of each odd-numbered strip you now have three $1/2''$ sections (Fig. 2–22a). Bend the sections on each end slightly forward; bend the center one slightly backward. Place coffee can rim up into these V-shaped troughs. Wrap the end sections back around rim and the center section forward around rim (Fig. 2–22b).

Form the strips into a bell shape as you attach each one to the rim. (Notice how the ridges on the can help to shape the bell.)

Paint the bell and hang a Christmas tree ball on a fine wire inside the bell for a clapper. Arrange several bells together with a spray of pine and a big red ribbon.

If you would like to make a smaller bell, use a $2^5/8''$-diameter can, $3^1/4''$ tall, with the rim from a 4″ can for bottom lip.

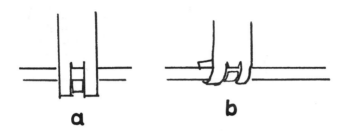

a   b

Fig. 2–22   Attaching sections of odd-numbered strips over rim

TRIVETS

To make the variety of trivets shown in Fig. 2–23, you'll need several long strips of tin, $1/4''$ wide, cut from a 1-gallon can.

Trivet 1

Take four strips of tin, each $12^1/2''$ long and $1/4''$ wide. Shape them as shown in Fig. 2–24a; clamp these strips together (Fig. 2–23, upper left).

Trivet 2

Take six strips of tin, each $9^1/2''$ long and $1/4''$ wide. Shape them as shown in Fig. 2–24b. Clamp them to a ring made from a strip of tin 8″ long (Fig. 2–23, upper right).

Trivet 3

Take six strips of tin, each 15″ long and $1/4''$ wide. Shape them as shown in Fig. 2–24c. Clamp them together (Fig. 2–23, bottom).

Fig. 2–23   Three styles of trivets

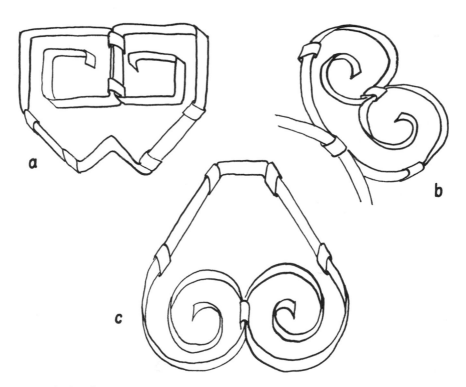

Fig. 2–24   Patterns for bending trivet sections

Chapter 3

Flowers and Bouquets

All the flowers in this chapter, shown in Fig. 3–1 and in color Fig. 1, are made from either aluminum or tin soda or beer cans. If aluminum cans can be found in your area, use them for the small flowers: They are easier to cut and shape. If aluminum cans are not available, use tin.

Fig. 3–1   A bouquet of flowers

CAN PREPARATION AND MATERIALS

The first step is to remove the top and bottom rims from cans. Cut the seam out if you're using tin, lay sheet of metal flat and draw circles (Fig. 3–2).

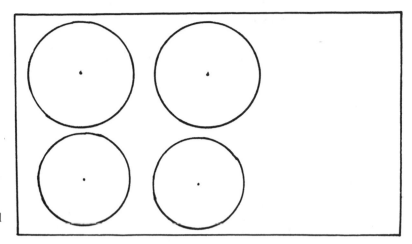

Fig. 3–2 Aluminum can prepared to make flowers

All the flowers are bolted together using a 6 x 32 x $\frac{1}{2}''$ bolt. These bolts are readily available at any hardware store (Fig. 3–3). For stems I use a small coil of 18-gauge wire, purchased at the hardware store. The stems are all bolted on the same way. Just bend a small loop in one end of the wire and bolt it on (Fig. 3–3). When wrapped with floral tape, the nut and bolt appear as the calyx of the flower.

In order to give your flowers the natural look, I would suggest you use seed packets or flower catalogues for color guides. The flowers are painted with acrylics and a brush or are sprayed with craft matte-finish paints before they are assembled.

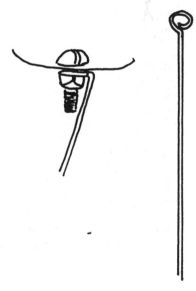

Fig. 3–3   Bolting blossoms to stems

The directions for individual flowers call for scoring with an awl; this scoring is just a mark or indentation in the metal (Fig. 3–4a). *Do not* press too hard, as excess pressure will cut the thin metal.

The soda or beer can labels are always painted on the outside of the can. As you work with these cans, always keep the painted side down. Thus, when the directions call for scoring on the under or painted side, it means the original paint on the can and not the finished paint you apply.

As an aid to guide you in making equal divisions when cutting your disks, simply cut out cardboard circles a little larger than the metal disks you will be using. Mark these circles as shown in Fig. 3–4.

Materials needed to make the flowers in this chapter are: cans, cans, and more cans; wire for stems; 6 x 32 x ½″ bolts; floral tape; paint.

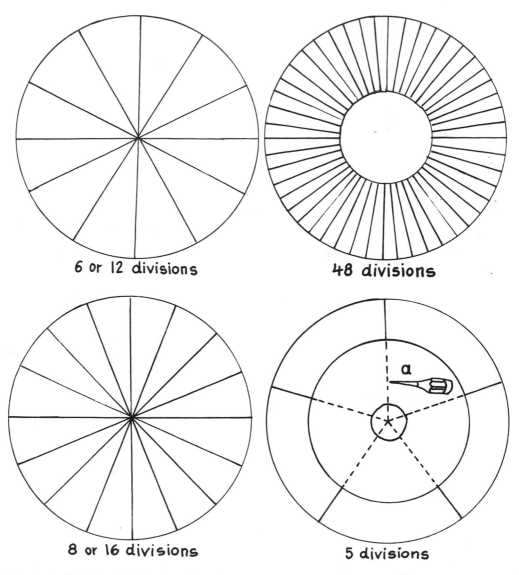

Fig. 3–4   Dividing disks into equal sections

LEAVES

To attach leaves to stems of flowers, first wrap flower stem with floral tape. Then with needle-nose pliers, bend a crease in the stem of the leaf (Fig. 3–5a). Fold leaf stem around flower stem (Fig. 3–5b). Then wrap with floral tape again.

If you intend to make a whole basket of flowers, why not skip the individual leaves? Make several of the large fern-type leaves to put in the basket as filler (Fig. 3–7a).

The dimensions I give for the various flowers are flexible. I find the dimensions I use suit my taste. After you have made a few flowers, try making them either larger or smaller; they may serve your purposes better.

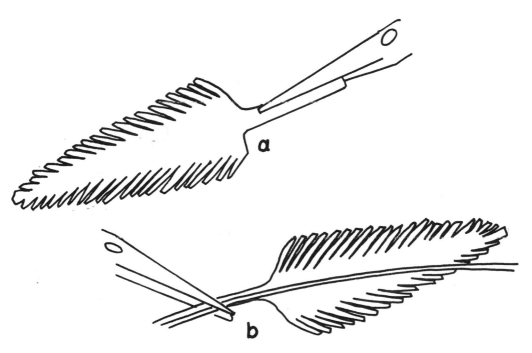

Fig. 3–5   Attaching leaves to stems

ASTERS

To make the aster shown in Fig. 3–8, cut five disks: two $1^7/_8''$, one $1^5/_8''$, one $1^1/_4''$, and one $^3/_4''$ in diameter.

Draw a $^1/_2''$ circle in the center of each disk (Fig. 3–9a). Punch a hole in the center of each disk just large enough for a 6 x 32 x $^1/_2''$ bolt.

Divide each disk into sixteen equal sections and cut to the $^1/_2''$ circle (Fig. 3–9b). Cut the sixteen sections into the petal shapes (Fig. 3–9c).

With an awl or icepick, score each petal. Score from the top or unpainted side of the disk as indicated by the dotted line in Fig. 3–9d. Paint and set aside to dry.

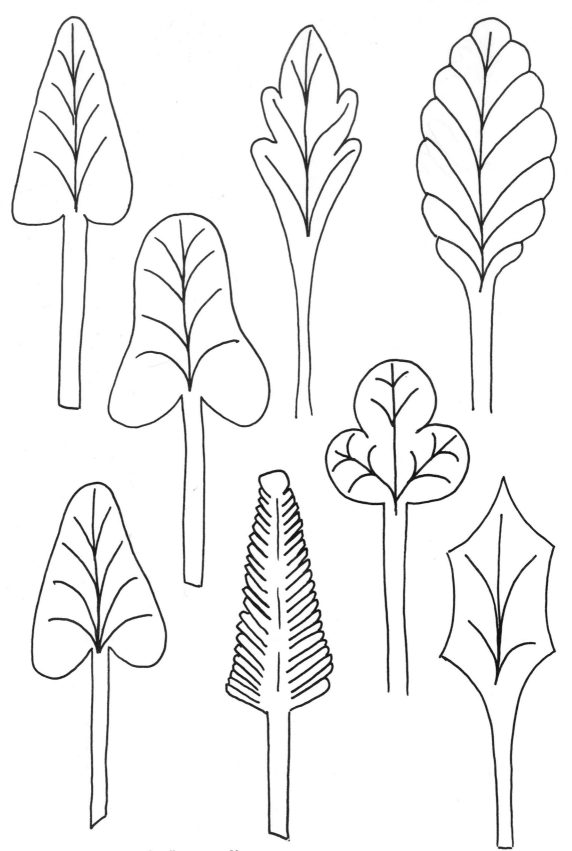

Fig. 3–6   Patterns for all manner of leaves

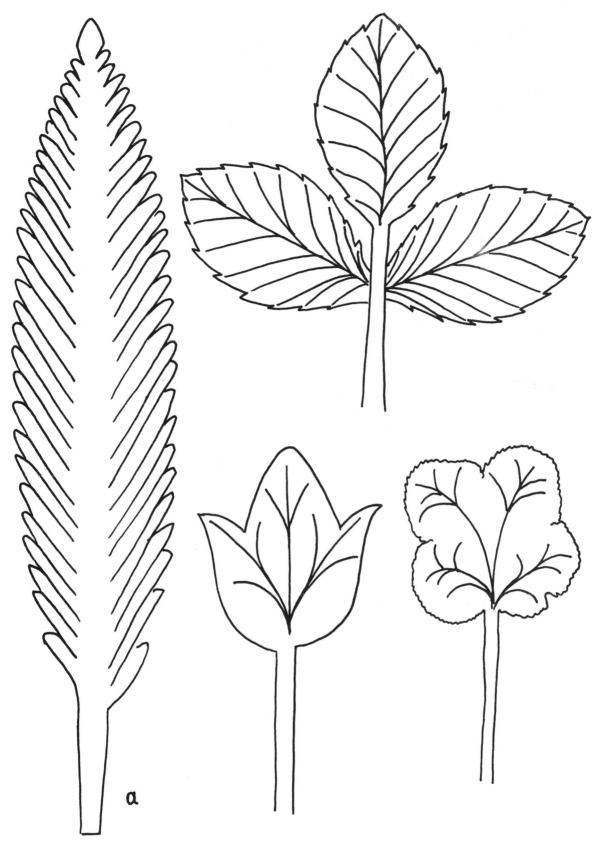

Fig. 3–7   More leaf patterns

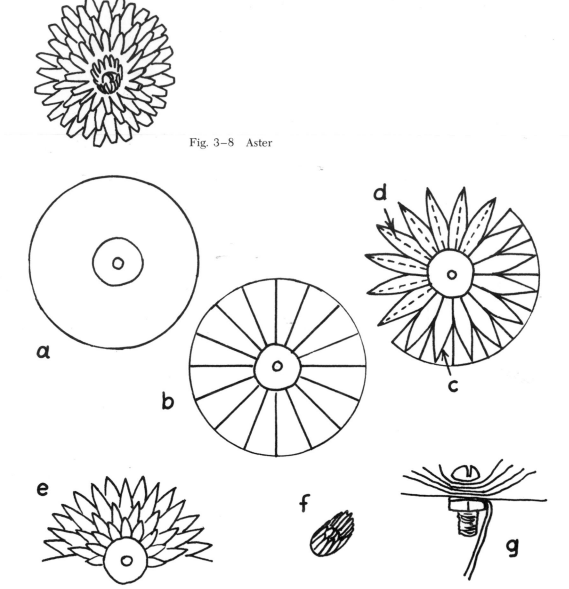

Fig. 3–8   Aster

Fig. 3–9   Step-by-step layout for making asters

Bolt together, putting small disk on top and working down to the largest ones, then adding the wire. As you tighten the bolt, make sure to keep the petals in a staggered position (Fig. 3–9e).

After the disks are bolted together as tightly as possible, form the petals of the center disk up and over the head of the bolt (Fig. 3–9f). This center piece should be painted yellow and may need to be touched up now.

The remaining four disks should be formed as shown in profile drawing (Fig. 3–9g). With floral tape, wrap bolt and wire stem to form calyx of the flower. Add leaves and wrap again.

DAISIES OR BROWN-EYED SUSANS

Cut two disks, each 1⁷/₈″ in diameter, to make the flower in Fig. 3–10. Draw a ¹/₂″ circle in the center of each disk (Fig. 3–11a). Punch a hole in center of each disk just large enough for a 6 x 32 bolt (Fig. 3–11b).

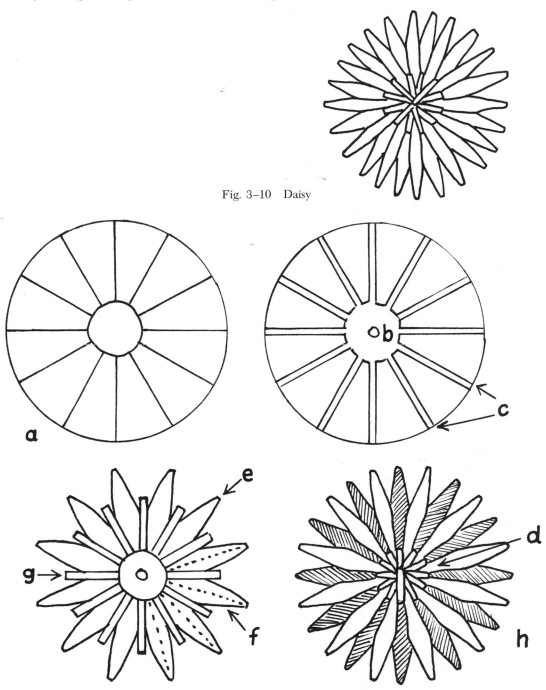

Fig. 3–10   Daisy

Fig. 3–11   Layout for making a daisy

Divide each disk into twelve equal sections (Fig. 3–11a), and cut to the $^1/_2''$ circle. Make another cut, about $^1/_{16}''$ wide, alongside each of the first twelve cuts (Fig. 3–11c).

On disk 1, which will be the top disk on the completed daisy, *do not* remove those twelve strips. On disk 2, which will be the bottom disk on completed daisy, *do* remove them. In the final assembly stage, these strips will fold over and form the center disk of the flower, hiding the head of the bolt (Fig. 3–11d).

Cut each of the twelve sections to the shape of petals (Fig. 3–11e).

With an awl or icepick, score each petal; score from the top or unpainted side of the disk (Fig. 3–11f, *see* dotted lines).

Snip off about $^1/_4''$ of each of the $^1/_{16}''$ strips on disk 1 (Fig. 3–11g).

Paint your flower now, before you assemble it. Let dry. Bolt flower together with stem wire on bottom.

The two disks should be bolted so the petals are staggered (Fig. 3–11h). Bend the $^1/_{16}''$ strips over the center of flower, hiding the head of the bolt (Fig. 3–11d).

This center should be painted yellow and may need to be touched up a bit now.

With floral tape, wrap stem and bolt to form calyx of the flower. Add leaves and wrap again.

If making brown-eyed Susans, paint petals yellow and center brown.

CORNFLOWERS OR BACHELOR BUTTONS

Cut four disks: two $1^1/_2''$, one $1^1/_4''$, and one $1''$ in diameter. Cut out the three larger disks using an old pair of pinking shears. Cut outside the measurement mark (Fig. 3–13a). After cutting around once, go around each disk again with pinking shears as shown in Figs. 3–14 and 3–15. This will give the scalloped edge a finer look.

Draw a $^1/_2''$ circle in the center of each disk (Fig. 3–13a). Punch hole in center of each disk just large enough for a 6 x 32 bolt.

Divide the three larger disks into eight sections (Fig. 3–13b) and cut to the $^1/_2''$ circle. After you have made the cuts to the $^1/_2''$ circle, widen these cuts as shown in Fig. 3–13c.

With needle-nose pliers, crimp each petal of the three larger disks as shown in Fig. 3–16. Don't try to be too methodical in this operation—it should have a natural, ragged look.

Fig. 3–12   Bachelor button

Draw a $\frac{1}{2}''$ circle in the center of the 1″ disk. Punch a hole for the bolt.

Divide and cut this disk into sixteen sections. Cut each section to the shape shown in Fig. 3–13d. These little petals will be the center of the finished flower. They will fold up, hiding the head of the bolt.

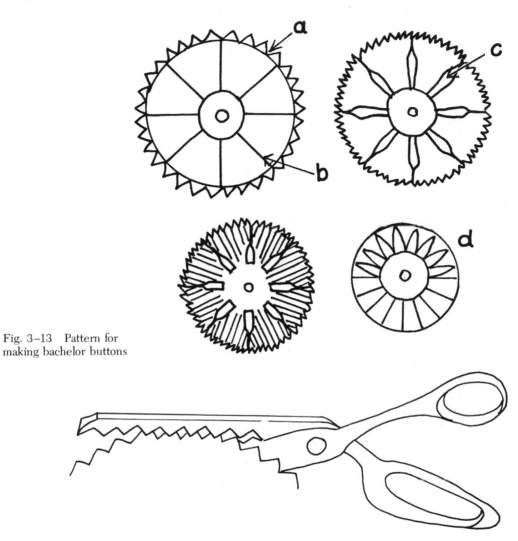

Fig. 3–13   Pattern for making bachelor buttons

Fig. 3–14   Cutting the edges with pinking shears

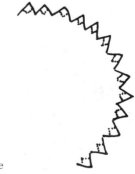

Fig. 3–15   Cutting again for a finer edge

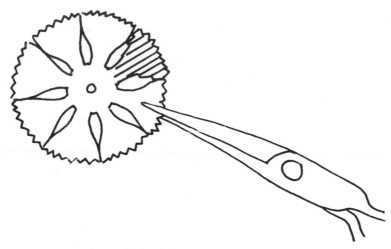

Fig. 3–16   Crimping the bachelor button

Paint. Set aside to dry.

Bolt together, putting the small disk on top and larger ones on the bottom, followed by stem. Wrap with floral tape. Add leaves and wrap again.

PINKS

To make the flower shown in Fig. 3–17, cut four disks: one $1^1/_2''$, two $1^1/_4''$, and one $1''$ in diameter.

Cut out three of the disks—one of each size—with pinking shears, cutting to the outside of the marked circles. After cutting out the disks, go around each one again with pinking shears. This will give the scalloped edge a finer look (*see* Figs. 3–14 and 3–15).

Draw a $^1/_2''$ circle in the center of each disk (Fig. 3–18a). Punch a hole in the center of each disk just large enough for a 6 x 32 bolt (Fig. 3–18b).

Divide and cut these three disks into sixteen sections; cut to the $^1/_2''$ circle (Fig. 3–18c). It does not matter if these sections are equal.

Crimp each section with needle-nose pliers (Fig. 3–18d), following directions for making cornflowers.

Paint pink and allow to dry.

Fig. 3–17   Pinks

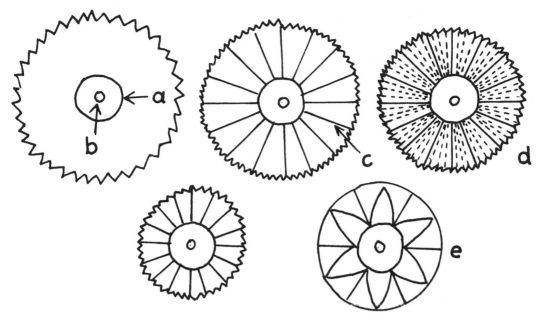

Fig. 3–18   Layout for making pinks

With regular snips, cut out the other 1¼″ disk. Mark a ½″ circle in the center and punch for bolt. Divide into six equal sections and cut to the ½″ circle (Fig. 3–18e).

Cut to shape as shown in Fig. 3–18e. This will form the sepal on the pink. Paint green.

Bolt together, starting with small disk on top and adding each larger disk in succession, ending with the sepal and stem wire. Bend all disks up tight. Wrap with floral tape, add leaves, and wrap again.

ROSES

To make a rose, as shown in Fig. 3–19, cut four disks: one 1¾″, one 1½″, and two 1¼″ in diameter.

Draw a ½″ circle in the center of each disk (Fig. 3–20a). Punch a hole in the center of each disk just large enough for a 6 x 32 bolt (Fig. 3–20b).

Fig. 3–19   Rose

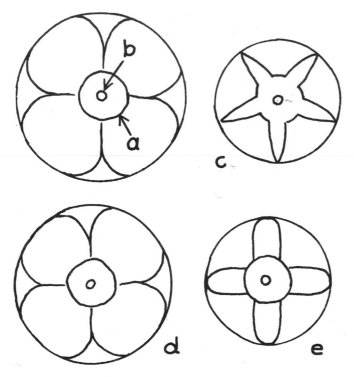

Fig. 3–20   Design for making roses

Divide one of the 1¼″ disks into five equal sections (Fig. 3–20c) and cut to shape as shown. Paint this disk dark green. It is the sepal of the rose. Set it aside for the time being.

Divide the other three disks into four equal sections and cut to the ½″ circle. Cut the four sections into the shape of petals as shown in Fig. 3–20d. Shape the smaller of these three disks (1¼″) as shown in Fig. 3–20e. Lay aside.

Fold a heavy piece of material, such as a bath towel, into several thick-

Fig. 3–21   Shaping rose petals with ball-peen hammer

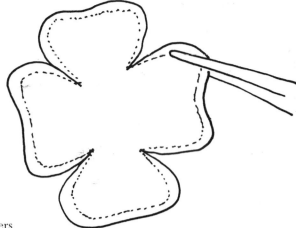

Fig. 3–22   Shaping rose petals with pliers

nesses. Lay this on a sturdy table. Lay the disks on this soft mat and tap each of the four petals very hard with the ball or peen end of the hammer (Fig. 3–21). This tapping action tends to stretch the metal into a concave or cup shape.

After the two larger disks have been peened out, bend about a $^1/_{16}''$ lip around each of the four petals on both disks (Fig. 3–22). Use the needle-nose pliers for this.

Paint. Set aside to dry. Bolt together, starting with the smaller $1^1/_4''$ disk, then the two larger disks, then adding the sepal and stem. As you tighten the bolt, be sure to keep the petals in a staggered position.

Wrap stem with floral tape, add leaves, and wrap again.

LILLIPUT ZINNIAS

To make the Lilliput zinnia shown in Fig. 3–23, cut two disks: one $1^5/_8''$ and one $1^1/_2''$ in diameter.

Draw a $^1/_2''$ circle in the center of each disk (Fig. 3–24a). Punch a hole in center of each disk just large enough for a 6 x 32 bolt (Fig. 3–24b).

Divide each disk into eight equal sections and cut to the $^1/_2''$ circle (Fig. 3–24c).

Fig. 3–23   Lilliput Zinnia

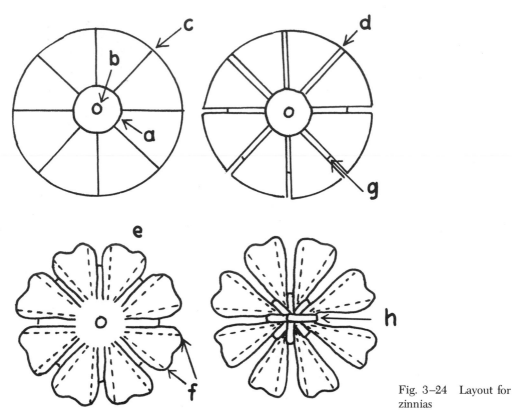

Fig. 3–24   Layout for zinnias

Cut a strip about $1/16''$ wide down the length of each section (Fig. 3–24d). In the final assembly stage, these eight strips will form the center of the zinnia and provide a spacer between the two disks.

Cut the eight sections to the shape of petals (Fig. 3–24e). With an awl or icepick, score each petal twice as shown in Fig. 3–24f. Score from under or painted side along dotted lines.

On the larger disk ($1^5/8''$), snip off about $1/4''$ of the eight strips (Fig. 3–24g). Fold the remaining part in toward the center of disk. This will give a little space between the two disks when they are finally bolted together. Paint. Set aside to dry.

Bolt together, putting the smaller disk on top and stem wire under the larger disk. As you tighten the bolt, make sure petals are staggered.

After the disks are bolted together as tightly as possible, form the eight narrow strips of the top disk up and over the head of the bolt (Fig. 3–24h). This center should be painted yellow.

Wrap the bolt and stem with floral tape. Add leaves and wrap again.

MERRIGOLDS

To make the merrigold in Fig. 3–25, cut five disks: one $1^5/8''$, one $1^1/2''$, one $1^3/8''$, one $1''$, and one $3/4''$ in diameter.

Fig. 3–25   Merrigold

Draw a $1/2''$ circle in the center of each disk (Fig. 3–26a). Punch a hole in center of each disk large enough for a 6 x 32 bolt (Fig. 3–26b).

Divide each of the four largest disks into eight equal sections (Fig. 3–26c) and cut to the $1/2''$ circle.

Cut the eight sections, on all four disks, to the shape of petals (Fig. 3–26d).

Using your needle-nose pliers as a guide, bend both edges of each petal up (Fig. 3–27).

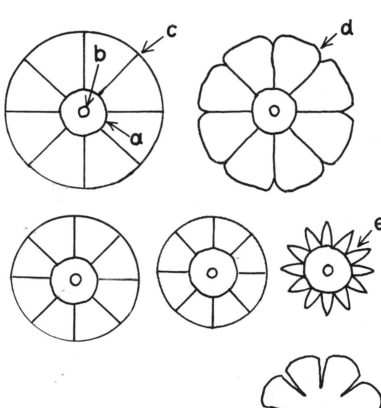

Fig. 3–26   Sequence of steps for making merrigolds

Fig. 3–27   Shaping merrigold petals

Divide the 3/4″ disk into twelve equal sections (Fig. 3–26e), and cut to the 1/2″ circle. Cut these twelve sections into petal shapes (Fig. 3–26e).

Paint the top side or cup of each petal orange. Paint the bottom of each petal yellow. Set aside to dry. Paint the small disk yellow on both sides.

Bolt together, putting small disk on top and working down to the largest disk and then the wire stem. As you tighten the bolt, make sure to keep all petals in a staggered position to the petals below them (Fig. 3–28a and c).

After the disks are bolted together as tightly as possible, form the petals of the center disk up and over the head of the bolt (Fig. 3–28b). This center piece should be painted yellow and may need some touching up now.

With floral tape, wrap the wire and bolt to form calyx of the flower. Add leaves and wrap again.

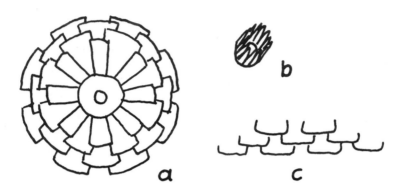

Fig. 3–28   Assembling merrigolds

SMALL MUMS

To make the flower in Fig. 3–29, cut three disks: one 1⅞″, one 1⅜″, and one 3/4″ in diameter.

Draw a 1/2″ circle in the center of each disk (Fig. 3–30a). Punch a hole in the center of each disk just large enough for a 6 x 32 bolt (Fig. 3–30b). Divide each of the three disks into sixteen sections and cut down to 1/2″ circle (Fig. 3–30c).

On the two larger disks, slit each of the sixteen sections in half (Fig. 3–30d), cutting only about three-quarters of the way down.

On the largest disk, slit each of the thirty-two sections in half (Fig. 3–30e), cutting only about halfway down. Paint; set aside to dry.

Bolt together, starting with the small disk and adding the larger disks in succession, ending with the wire stem.

After the mum has been assembled, bend some of the little sections down,

Fig. 3–29   Small chrysanthemum

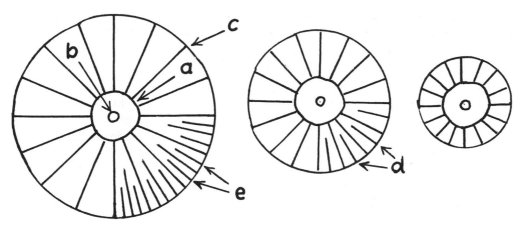

Fig. 3–30 Layout for making a mum

bend some up, bend some in, and bend some out. Get the idea? Make it look ragged.

AFRICAN VIOLETS

These flowers are perfect for hanging baskets. Cut a sheet of aluminum from a can. Using pattern for violets (Fig. 3–31a), draw as many as you will need to fill a basket. Cut and punch two small holes for wire stems (Fig. 3–31b). Take a 6″ length of 28-gauge wire, bend it in half, and insert it into violet (Fig. 3–31c). Wrap with floral tape. Paint.

Fig. 3–31 Making bouquets of violets

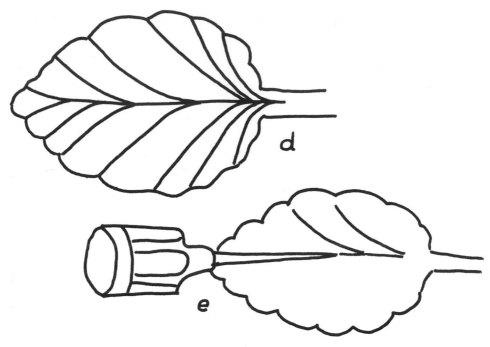

Fig. 3–31, continued Making bouquets of violets

Fig. 3–31d gives you the pattern for violet leaves. Cut leaves out of another sheet of aluminum. Mark the veins of the leaves by placing the leaf on a magazine or several thicknesses of newspaper, which gives a semisoft pad to work on. Etch the lines into the surface of the leaves with the awl or icepick (Fig. 3–31e). Be careful not to press too hard—just enough to indent, but not to cut through.

SMALL MIXED FLOWERS

For the mixed flowers, draw circles the size of a quarter with your compass. Cut these circles into the shapes shown in Fig. 3–32. Put wire stems on as de-

Fig. 3–32   Sprigs of small flowers

scribed for violets. Twist three or four flower stems together, then cover with floral tape.

The greens or fillers for small flowers are very easy to make. Lay a sheet of aluminum flat, mark 1″ up from bottom (Fig. 3–33a). Cut into 1½″ sections. Cut about ¹/₁₆″-wide strips down to the 1″ mark. Paint green. Let dry. Now roll up as shown (Fig. 3–33b); then fan out.

Put several of these in the basket with flowers. Insert flowers and leaves into a small glob of floral clay in bottom of basket.

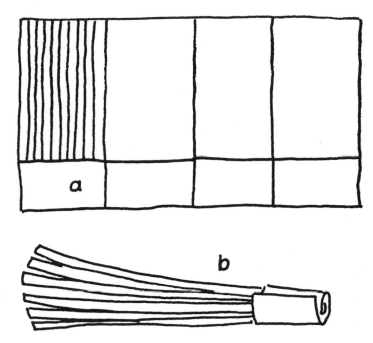

Fig. 3–33   Sprays of green filler for small bouquets

Wouldn't this be a much prettier country if all the cans along our highways could be transformed into beautiful flowers? Get out your magic wand (tin snips) and go to work. If you make all the flowers in this book, at least fifteen cans will have been recycled.

Let's all join in the fun!

Chapter  4
Dollhouse Furniture

HIGHCHAIR

To make the highchair shown in Fig. 4–1 and in the color section (Fig. 7), you'll need these materials: one tall $2^1/8''$-diameter can; one lid from a dry mustard spice can; one beverage can, any size (you will need a flat piece of tin from the beverage can for a foot rest and several extra strips).

Remove top and top rim. Remove side seam. Divide can into eight equal sections (*see* Ch. 1, Working with Cans). Divide each section into eight equal strips.

Hold the can bottom-side-up and number each strip from 1 through 64, starting to the left of the space where seam was removed (marked by an asterisk in Fig. 4–2). Work from right to left.

Fig. 4–1   Cradle, stroller, and highchair

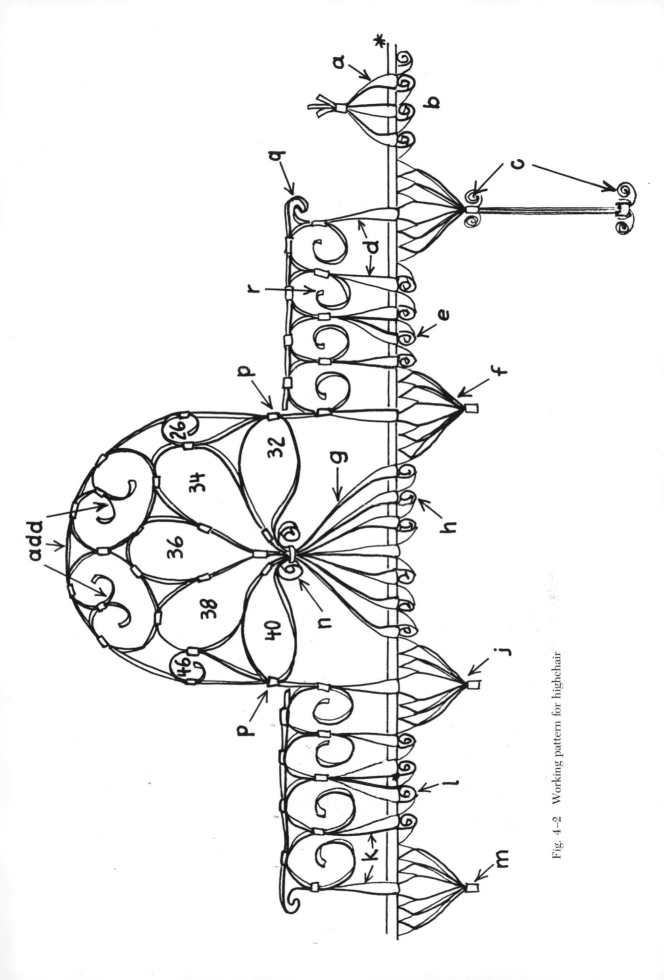

Fig. 4-2　Working pattern for highchair

Bend even-numbered strips, 2, 4, and 6, up and leave for the time being (Fig. 4–2a). Shorten strips 1, 3, 5, and 7 to about $1^1/_2''$ from bottom of can. Twist them one quarter turn counterclockwise. Curl them up tight toward the right (Fig. 4–2b).

Twist strips 8, 9, and 10 one-quarter turn clockwise. Bend strip 11 up and leave. Twist strips 12, 13, and 14 one-quarter turn counterclockwise. Clamp strips 8, 9, 10, 12, 13, and 14 together about $1''$ below base of can. Snip strips 8 and 14 off $2''$ below clamp and curl them back over the clamp. Place another clamp about $3''$ or $3^1/_2''$ below base of can. Snip strips 9 and 13 off about $1/_8''$ below this clamp and bend them back over the clamp. Curl strips 10 and 12 up over the clamp (Fig. 4–2c). This will form leg number one.

Bend strips 16, 18, 20, and 22 up and leave for the time being—these will form side of chair (Fig. 4–2d). Shorten strips 15, 17, 19, and 21 to about $1^1/_2''$ from base of can. Twist them one-quarter turn clockwise and curl them up tight toward the left (Fig. 4–2e).

Twist strips 23, 24, and 25 one-quarter turn clockwise. Bend strip 26 up and leave. Twist strips 27, 28, and 29 one-quarter turn counterclockwise. Clamp strips 23, 24, 25, 27, 28, and 29 together about $1''$ below base of can. Snip strips 23 and 29 off $2''$ below clamp and curl them up over the clamp. Place another clamp about $3''$ or $3^1/_2''$ below base of can. Snip strips 24 and 28 off $1/_8''$ below this clamp and bend them back over the clamp. Curl strips 25 and 27 up over the clamp (Fig. 4–2c). This will form the second leg (Fig. 4–2f).

Bend strips 30, 32, 34, and 36 up and twist them one-quarter turn clockwise. Bend strips 38, 40, and 42 up and twist them one-quarter turn counterclockwise. Clamp these seven strips together about $1''$ from bottom of can. These, along with strips 26 and 46, will form the back of chair (Fig. 4–2g).

Shorten strips 31, 33, 35, 37, 39, and 41 to about $1^1/_2''$ from bottom of can. Twist the first three one-quarter turn counterclockwise and curl them up tight toward the right. Twist the other three one-quarter turn clockwise and curl them toward the left (Fig. 4–2h).

Twist strips 43, 44, and 45 one-quarter turn clockwise. Bend strip 46 up and leave. Twist strips 47, 48, and 49 one-quarter turn counterclockwise. Clamp strips 43, 44, 45, 47, 48, and 49 together about $1''$ below base of can. Snip strips 43 and 49 off about $2''$ below clamp and curl them up over the clamp. Place another clamp about $3''$ or $3^1/_2''$ below base of can. Snip strips 44 and 48 off $1/_8''$ below this clamp and bend them back over the clamp. Curl strips 45 and 47 up over the clamp (Fig. 4–2c). This will form the third leg (Fig. 4–2j).

Bend strips 50, 52, 54, and 56 up and leave for the time being—these will form the other side of chair (Fig. 4–2k).

Shorten strips 51, 53, 55, and 57 to about $1^1/_2''$ from bottom of can. Twist them one-quarter turn counterclockwise and curl them in tight toward the right (Fig. 4–2l).

Twist strips 58, 59, and 60 one-quarter turn clockwise. Bend strip 61 up and leave. Twist strips 62, 63, and 64 one-quarter turn counterclockwise. Clamp strips 58, 59, 60, 62, 63, and 64 together about $1''$ below base of can. Snip strips

58 and 64 off about 2″ below clamp and curl them up over the clamp. Place another clamp about 3″ or 3¹/₂″ below base of can. Snip strips 59 and 63 off ¹/₈″ below this clamp and bend them back over the clamp. Curl strips 60 and 62 up over the clamp (Fig. 4–2c). This will form the fourth leg (Fig. 4–2m).

Now finish the back of the highchair. Shorten strips 30 and 42 to about 1¹/₂″ from clamp and curl them in tight over the clamp (Fig. 4–2n). Shorten strips 32 and 40 to about 2″ from clamp. Leave strips 34, 36, and 38 their original length. Shape these up, along with strips 26 and 46, as shown in diagram.

Note: The design in back of chair will depend on the height of your 2¹/₈″ can. The can I used was 4″ tall, but I feel sure you can work out a similar design with a 3 or 3¹/₂″ can.

Cut three 5″ strips and add the end of one at the point where strips 40 and 46 are bound together (Fig. 4–2p). Clamp other end where strips 26 and 32 are clamped together. Curl the other two strips, as shown, and clamp them into place (Fig. 4–2 add).

Now finish the side. Place your curling tool about 1¹/₂″ above the bottom of can on strips 11 and 61 (Fig. 4–2q). Curl down one and a half revolutions of curling tool. Twist strips 16 and 18 one-quarter turn counterclockwise. Twist strips 54 and 56 one-quarter turn clockwise. Curl them and clamp them as shown (Fig. 4–2r).

Now finish the front tray. Cut two strips about 3″ long. Wrap them around the mustard top and clamp them real tight (Fig. 4–3a). Slip the ends through the arm uprights (strips 11 and 61) and clamp again (Fig. 4–3b). Twist strips 2 and 4 one-quarter turn clockwise. Twist strip 6 one-quarter turn counterclockwise. Clamp these three about ³/₈″ from bottom of can. Snip strips 2 and 6 off about ¹/₈″ above clamp. Then bend them down over the clamp. Slip strip 4 between the tray and the strip holding it. Snip strip 4 off about ¹/₈″ above tray and then bend it down (Fig. 4–3c).

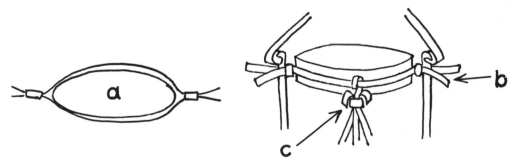

Fig. 4–3 Highchair tray

Cut a flat piece of tin as shown (Fig. 4–4a). Bend the ears on this piece down and around each leg just below the top clamp (Fig. 4–4b).

Paint. Add a seat if desired.

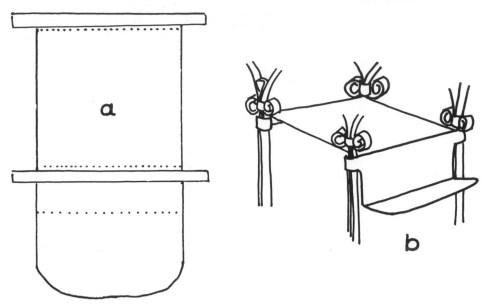

Fig. 4–4   Footrest for highchair

STROLLER

The stroller is shown in Fig. 4–1 and in Fig. 7 (color section). To make it, you'll need one $2^1/_8''$-diameter can; one set of four small wheels and axles; one 1-gallon oil can; a flat piece of tin for a footrest, plus a long narrow strip for the handle.

Remove top and top rim from small can. Remove side seam. Divide can into eight equal sections. Divide each section into eight equal strips (*see* Ch. 1). Hold the can bottom-side-up and number each strip from 1 through 64, starting at the left of space where seam was removed.

Twist strips 1, 2, and 3 one-quarter turn clockwise, as shown in Fig. 4–5. Bend strip 4 up and leave for the time being. Twist strips 5, 6, and 7 one-quarter turn counterclockwise. Clamp strips 1, 2, 3, 5, 6, and 7 together about $3/_4''$ from bottom of can (Fig. 4–5a). Shorten strips 1 and 7 to about $1^1/_2''$ from clamp and curl them up tight over clamp (Fig. 4–5a). This group will form a leg. Leave it for the moment.

Bend strips 8, 9, 10, and 11 straight up and twist them one-quarter turn clockwise. Remove strip 12. Bend strips 13, 14, 15, and 16 straight up and twist them one-quarter turn counterclockwise. Clamp these eight strips together about $3/_8''$ from base of can (Fig. 4–5b).

Snip strips 8 and 16 off about $1/_8''$ above this clamp. Bend them down tight over clamp. Put a clamp on the remaining six strips $3/_8''$ above first clamp (Fig. 4–5c).

Snip strips 9 and 15 off about $1/_8''$ above clamp. Bend them down tight over clamp. This group will form the front, or saddle, of stroller. We will come back to

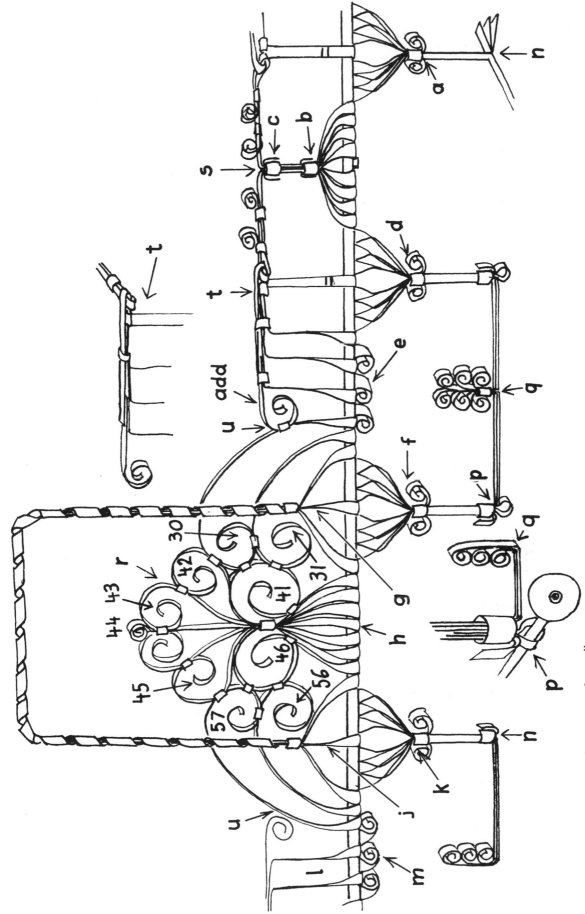

Fig. 4-5 Working pattern for stroller

them in a moment. (I wear a self-winding watch and, the way I move, the moments may be longer than they are on a conventional watch.)

Twist strips 17, 18, and 19 one-quarter turn clockwise. Bend 20 up and leave. Twist strips 21, 22, and 23 one-quarter turn counterclockwise. Clamp strips 17, 18, 19, 21, 22, and 23 together about $3/4''$ from bottom of can. Shorten strips 17 and 23 to about $1^1/2''$ from clamp, then curl them up tight over clamp. This group will form the second leg (Fig. 4–5d). Leave it for the time being.

Bend strips 24, 26, 28, 30, and 31 up, twisting each one-quarter turn counterclockwise; leave. Bend strip 32 up. Shorten strips 25, 27, and 29 to about $1^1/2''$ from bottom of can. Twist them one-quarter turn clockwise and curl them down tight toward the left (Fig. 4–5e). Twist strips 33, 34, and 35 one-quarter turn clockwise. Bend strip 36 up and forget. Twist strips 37, 38, and 39 one-quarter turn counterclockwise. Clamp strips 33, 34, 35, 37, 38, and 39 together about $3/4''$ from bottom of can. Shorten strips 33 and 39 to about $1^1/2''$ from clamp. Then curl them up tight over clamp. This group will form the third leg (Fig. 4–5f).

Bend strip 40 up and twist it one-quarter turn counterclockwise. Now go back and twist strips 36 and 32 one-quarter turn clockwise. Clamp strips 32, 36, and 40 together about $1/2''$ above bottom of can. This group will form one side of the stroller handle (Fig. 4–5g).

Bend strips 41, 42, 43, and 44 up and twist them one-quarter turn clockwise. Bend strips 45, 46, and 47 up and twist them one-quarter turn counterclockwise. Clamp these seven together about $3/4''$ above bottom of can. This group will form the back (Fig. 4–5h).

Bend strip 48 up and twist it one-quarter turn clockwise. Bend strips 52 and 56 up and twist them one-quarter turn counterclockwise. Clamp these three together about $1/2''$ above bottom of can. This group will form the other side of the stroller handle (Fig. 4–5j).

Twist strips 49, 50, and 51 one-quarter turn clockwise. Twist strips 53, 54, and 55 one-quarter turn counterclockwise. Clamp strips 49, 50, 51, 53, 54, and 55 together about $3/4''$ from bottom of can. Shorten strips 49 and 55 to about $1^1/2''$ from clamp, then curl them up tight over clamp (Fig. 4–5k). This group will form the fourth leg.

Bend strips 57, 58, 60, 62, and 64 up. Twist all of them one-quarter turn counterclockwise. This group will form the second side (Fig. 4–5l).

Shorten strips 59, 61, and 63 to about $1^1/2''$ from bottom of can. Twist them one-quarter turn counterclockwise and curl them down tight toward the right (Fig. 4–5m).

Now go back and finish up the four legs. Just below the curls on each leg, get a firm grip on the remaining four strips with your pliers. Turn them slightly until flat sides of strips face front and back. A set of four wheels, each about the diameter of a dime, and two axles removed from a ten-cent store toy will serve nicely for the stroller wheels.

At a point $3/4''$ below the clamps on the front legs, bend one strip on each leg forward (strips 6 and 18) and three strips on each leg backward (strips 2, 3, and 5; 19, 21, and 22), as shown in Fig. 4–5n. Cut four strips of tin, each $1^1/2''$ long, the same width as the legs (approximately $1/8''$). Fold these strips, one at a time, over the axles and insert them in the legs as shown (Fig. 4–5p). Put a clamp on the legs at the bend.

Snip strips 6 and 18 off $1/8''$ from clamp and bend up over clamp. At a point $3/4''$ below clamp on back of the legs, bend one strip on each leg back (strips 38 and 50) and three strips on each leg forward (strips 34, 35, and 37; 51, 53, and 54). Insert axles and wheels the same as for front legs, then clamp.

Snip strips 38 and 50 off $1/8''$ from clamp and bend up over clamp. Bend each group of three strips up, $5/8''$ from axle. Clamp front and back groups together and curl down tight (Fig. 4–5q).

Now finish the back. Use a small curling tool to curl strip 44 down about $1\frac{1}{2}$ turns or until the tip is approximately $2''$ from bottom of can. Looking very closely at the drawings, shape up the rest of back (Fig. 4–5r).

Now to the front. Bend strips 13 and 14 to the left and strips 10 and 11 to the right (Fig. 4–5s).

At a point $3/4''$ above bottom of can, bend strips 24, 26, and 28 and strips 60, 62, and 64 toward front of stroller. These strips should cross over the front strips. Fold these front strips and clamp (Fig. 4–5t).

Curl strips 10 and 14 back. Clamp again and curl strips 11 and 13 back. Fold strips 4 and 20 over the side strips. Shape them around the side strips and clamp them onto themselves near bottom of can. Fold the side strips back over the front strips and corner posts (4 and 20) and clamp. Slip an extra strip about $2''$ long into these groups and clamp again. These extra strips should curl back and clamp to the back of stroller (Fig. 4–5u).

Remove top, top rim, bottom, and bottom rim from a gallon oil can. Lay can out flat and cut a $1/8''$-wide strip off of the full length, approximately $18''$ long. Wrap this strip around a heavy piece of wire, as shown in Fig. 1–20. Remove wire from strip. Hold each end with pliers and stretch to the desired length for a handle. Put a bend in handle about $3''$ from each end. Slip one end of this coil down over strips 32, 36, and 40 and clamp. Slip other end over strips 48, 52, and 56 and clamp.

Now cut a flat piece of tin (from the gallon can) about $1\frac{1}{8}''$ wide and approximately $2''$ long. Fit over axles as shown for a footrest. Paint and then install a cushion if desired.

CRADLE

Materials needed for the cradle (Fig. 4–1 and color Fig. 7) are one 1-quart fuel or turpentine can (oblong $4\frac{1}{2}''$ x $2\frac{1}{4}''$ at bottom); one soup can, $2\frac{5}{8}''$ in diameter, for hood of cradle; several heavy $3/16''$-wide strips for rockers.

Remove top and top rim from the quart can. Remove side seam. Shorten can to about $4''$ from bottom.

Locate the centers of each of the four corners and cut a slit from top edge to the base of can at each point. Cut five strips, approximately $1/8''$ wide each, to each side of the corner cuts. (Pay particular attention to the corner close to the seam. You will need five strips between gap where seam was removed and the

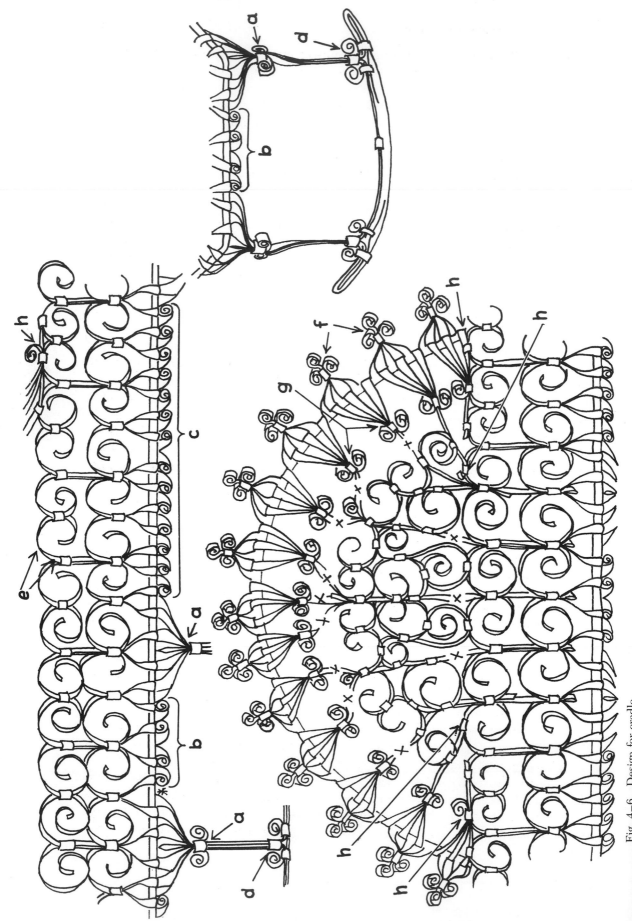

Fig. 4-6 Design for cradle

corner cut.) Temporarily clamp these groups of ten together (Fig. 4–6a). Six of them will form the legs. (The asterisk in Fig. 4–6 represents the spot where seam was removed.)

Divide the sections on each end between the legs into eight equal strips (Fig. 4–6b). Number these strips 1 through 8 (left to right). Bend strips 1, 3, 6, and 8 up. Snip strips 2, 4, 5, and 7 off to about $1^1/_2''$ from bottom of can. Twist strips 2 and 4 one-quarter turn clockwise and curl them up tight to the left. Twist strips 5 and 7 one-quarter turn counterclockwise and curl up tight to the right.

Divide the side panels between the legs into twenty-eight equal strips. To do this I would suggest that you divide the large panel into four smaller sections and then divide each of the four sections into seven strips each. Number these strips 1 through 28 (left to right). Bend strips 1, 3, 5, 7, 9, 11, 13, 16, 18, 20, 22, 24, 26, and 28 up. Snip strips 2, 4, 6, 8, 10, 12, and 14 off to about $1^1/_2''$ from bottom of can. Twist them one-quarter turn clockwise and curl them down tight to the left. Snip strips 15, 17, 19, 21, 23, 25, and 27 off to about $1^1/_2''$ from bottom of can. Twist them one-quarter turn counterclockwise and curl them down tight to the right (Fig. 4–6c).

Now go back to the four corners, remove clamps, number the strips 1 through 10 (left to right). Bend strips 2, 4, 7, and 9 up. Twist strips 1, 3, and 5 one-quarter turn counterclockwise. Twist strips 6, 8, and 10 one-quarter turn clockwise. Clamp these six strips together about $^5/_8''$ below bottom of can (Fig. 4–6a). Snip strips 1 and 10 off about $1^1/_2''$ from the clamp. Curl them down tight over the clamp. Place another clamp about 1″ below first clamp (Fig. 4–6d). Snip strips 3 and 8 off about $1^1/_2''$ from clamp and curl. (Don't tighten these curls up until rocker has been added.) Add rockers, following the procedure shown in Fig. 4–6.

You now should have fifty-two strips remaining in a standing position. Twist each of these strips one-quarter turn—one clockwise and the next counterclockwise—alternately, all the way around the can.

Now finish up the sides as shown in Fig. 4–6e. First clamp is about $^5/_8''$ from bottom and second clamp is $1^1/_2''$ from bottom.

To make the hood of the cradle, remove top and top rim from soup can. Remove side seam and bottom. Shorten can to about 3″ from bottom. Divide can into twelve equal sections. Divide each section into eight equal strips. This will give you a total of ninety-six strips.

Bend every other strip up over the rim of the can 180° to form the top of hood. Clamp remaining strips together in groups of three (Fig. 4–6f), about $^3/_8''$ from rim of can. Snip each of these strips off about $1^1/_2''$ from the clamp and curl them down tight over the clamp.

Clamp the strips that were bent up over the rim together in groups of three. Clamp them about $^5/_8''$ from rim. Snip the two outside strips, in each group, off to about $1^1/_2''$ from the clamp. Curl these down tight over the clamp (Fig. 4–6g).

Cut the rim in half at the spot where the seam was removed. Shorten rim to about $6^1/_2''$. Open rim out until the ends fit onto sides of cradle. Fasten as shown in drawing (Fig. 4–6h). Strips will have to have a twist or two at places marked by Xs on drawing. These twists help to shape the hood. Paint; add a mattress if desired.

PLAYPEN

Materials needed for this project (color section, Fig. 7) are one square can, $4^3/_4''$ x $4^3/_4''$ (salted cracker tin); several long $^1/_8''$-wide strips; fine wire

Remove top and top rim from can. Shorten to 5" from bottom of can; use excess material to cut the $^1/_8''$-wide strips. Remove side seam.

Locate the four corners and cut a slit from top edge down to the bottom of can. Cut five strips ($^1/_8''$ wide each) on each side of the corner cuts. Temporarily clamp these groups of ten together; six of them will form the legs.

Divide each of the large panels, between the legs, into three equal sections. Divide each of these sections into eight equal strips. You should now have twenty-four strips between each leg (Fig. 4–7a). Number them 1 through 24 (left to right). Remove strips 3, 7, 14, 18, and 22 (Fig. 4–7b). Bend strips 1, 5, 9, 12, 16, 20, and 24 straight up.

Shorten strips 2, 4, 6, 8, and 10 to about $1^1/_2''$ from base of can (Fig. 4–7d). Twist them one-quarter turn clockwise and curl them down tightly to the left.

Snip strips 11, 13, 15, 17, 19, 21, and 23 off about $1^1/_2''$ from base of can. Twist them one-quarter turn counterclockwise and curl them down tight to the right (Fig. 4–7e). Follow this same procedure on all four sides.

Now go back to the four corners and remove the temporary clamps. Number the strips in each group 1 through 10 (left to right). Remove strips 2 and 9 (Fig. 4–7f); bend strips 4 and 7 straight up; twist strips 1, 3, and 5 one-quarter turn counterclockwise. Twist strips 6, 8, and 10 one-quarter turn clockwise. Clamp these six together about $^1/_2''$ below bottom of can. Snip them off to about $1^1/_2''$ from clamp and curl down to form leg (Fig. 4–7g).

Referring back to Fig. 1–20, twist each of the thirty-six standing strips around a coathanger wire to form the slats in the side of the playpen. Finish each twisted slat off $2^1/_2''$ above bottom of can.

Cut several 12" to 18"-each long strips, $^1/_8''$ wide. Clamp the ends of each of the thirty-six twisted slats to one of these strips (see Fig. 4–7h). Twist several other long strips around a coathanger wire. Using a very fine wire, attach these long coils to top of playpen (Fig. 4–7j).

On one side of playpen, shorten two of the twisted slats to about $1^1/_2''$ (Fig. 4–7k). Insert three wires with beads, as shown in Fig. 4–7. Paint. Add a pad if desired.

ROCKING CHAIR

Materials needed for this project (Fig. 4–8) are one can, $3^1/_8''$ diameter and 5" tall (Campbell's Chunky Soup); two $^3/_{16}''$-wide strips of tin for rockers (use heavy gauge tin, such as a coffee can).

Remove top and top rim. Remove side seam. Divide into eight sections (see Ch. 1). Divide each section into eight strips. Hold the can bottom-side-up and number the strips 1 through 64. Starting to the left of space where seam was removed, count to your left.

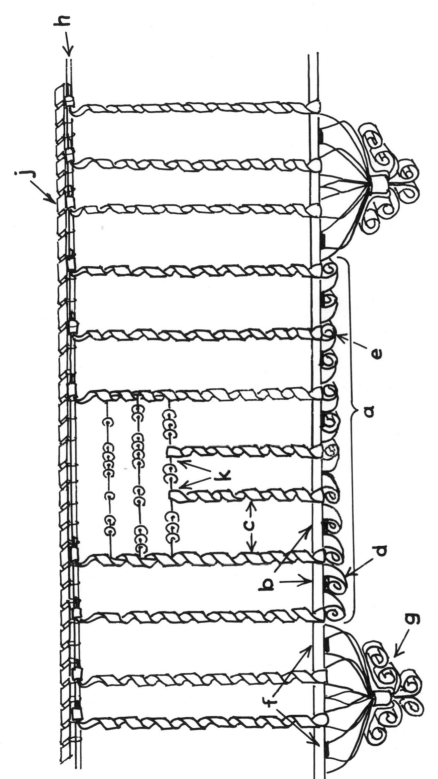

Fig. 4–7 Layout for working up the playpen

Fig. 4–8   Rocking chair

Remove strips 2, 4, 6, 8, and 10 (Fig. 4–9a). Do not throw these strips away. You will need them to make clamps.

Shorten strips 1, 3, 5, 7, 9, and 11 to about 1½″ below base of can. Twist them one-quarter turn counterclockwise and curl them up tight to the right (Fig. 4–9b).

Twist strips 12, 13, 14, and 15 one-quarter turn clockwise. Twist strips 16, 17, and 18 one-quarter turn counterclockwise. Clamp these seven strips together 1″ below bottom of can (Fig. 4–9c). This group (and what a fine group it is) will form one of the front legs. Leave it for the time being; you will come back to finish all four legs and the rockers later.

Bend strips 19, 21, 24, 26, and 30 straight up. Twist them one-quarter turn counterclockwise. This group of five will form one of the chair arms (Fig. 4–9d).

Shorten strips 20 and 22 to about 1½″. Twist them one-quarter turn counterclockwise and curl them up tight to the right. Shorten strips 23 and 25 to about 1½″. Twist them one-quarter turn clockwise and curl them up tight to the left (Fig. 4–9e).

Twist strips 27, 28, and 29 one-quarter turn clockwise. Twist strips 31, 32, and 33 one-quarter turn counterclockwise. Clamp these six strips together 1″ below bottom of can. This group will form one of the back legs (Fig. 4–9f).

Twist strips 34, 35, 36, and 37 one-quarter turn counterclockwise. Twist strips 38, 39, 40, 41, and 42 one-quarter turn clockwise. Clamp these nine strips together about 1″ above bottom of can. This group will form the back (Fig. 4–9g).

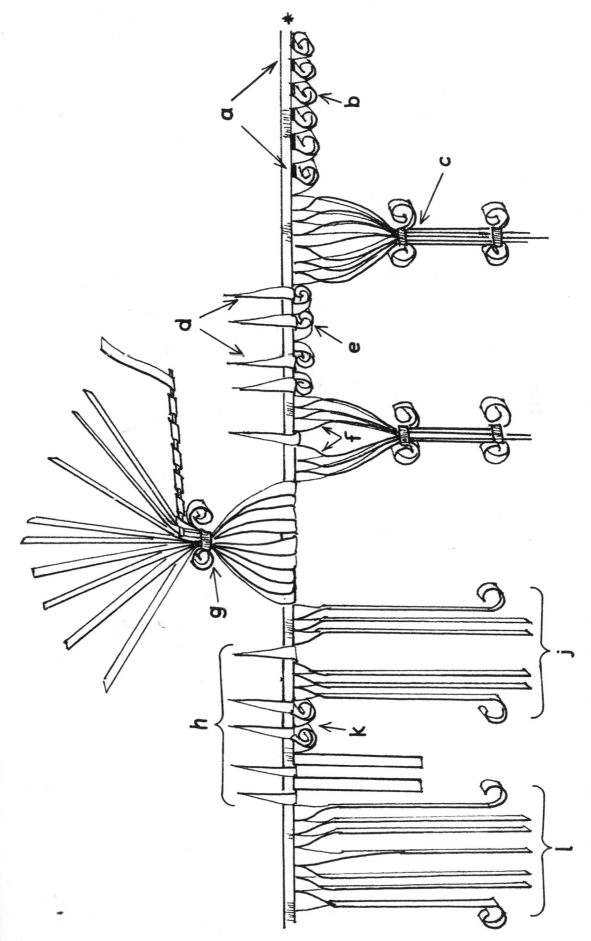

Fig. 4-9 Working pattern for the rocking chair

Fig. 4–10 Design for back and arms of rocking chair

Bend strips 46, 50, 52, 55, and 57 straight up. Twist them one-quarter turn clockwise. This group of five will form the second arm (Fig. 4–9h).

Twist strips 43, 44, and 45 one-quarter turn clockwise. Twist strips 47, 48, and 49 one-quarter turn counterclockwise. Clamp these six together 1″ below bottom of can. This group will form the second back leg (Fig. 4–9j).

Shorten strips 51 and 53 to about 1½″, twist them one-quarter turn counterclockwise and curl them up tight to the right. Shorten strips 54 and 56 to about 1½″, twist them one-quarter turn clockwise and curl them up tight to the left (Fig. 4–9k).

Twist strips 58, 59, 60, and 61 one-quarter turn clockwise. Twist strips 62, 63, and 64 one-quarter turn counterclockwise. Clamp these seven strips together 1″ below bottom of can. This group will form the fourth leg (Fig. 4–9l).

Strips 34 through 42 were clamped together earlier to use in forming the back. Now go back and finish this section. Shorten the two outside strips, 34 and 42, to about 1½″ from the clamp. Curl them down tight over the clamp (Fig. 4–9g). Slit the remaining seven strips down the middle. You now have fourteen narrow strips. Since you only want ten, look at the diagram very closely and see how the unwanted four are removed. Using the technique shown in Fig. 1–20, wrap these ten strips around a coathanger wire. Leave about 1½″ on the end of each strip uncoiled.

Form the back and the two arms as shown in Fig. 4–10.

Now you are ready to finish the four legs and add the rockers. Check to make sure the clamps on each of the four legs haven't moved. Each one should be 1″ below the bottom of can. Shorten the two outside strips on each leg to about 1½″ from clamp and curl them up tight over the clamps (strips 12 and 18; 27 and 33; 43 and 49; 58 and 64). Place another clamp 1″ below the first. Again shorten the two outside strips on each leg and partially curl them up. Do not tighten these curls until rocker has been installed (strips 13 and 17; 28 and 32; 44 and 48; 59 and 63).

Form rocker and install as shown in Fig. 4–6. Paint.

Add a cushion as shown in plans for footstool (Fig. 2–18).

FLOOR LAMP

Materials needed for this project are one lid or bottom from a 1-pound coffee can; one lid or bottom from a 1-gallon round can; one cigar tube; 10 inches of 16-gauge wire; two 6 x 32 x ½″ bolts.

The base of the lamp will be made from the 1-pound can lid. Find center of disk. Place one point of your compass in center, place other point of compass 1⅜″ out from center. Draw a circle and cut out. The 2¾″ disk you just cut out will be the base for the floor lamp. If you choose the correct can and cut it in the right place, the center 2″ of this disk should be high enough off the table so that the head of the bolt will not touch the table when base of cigar tube is attached (Fig. 4–11a).

Punch a hole in the center of this disk just large enough to take a 6 x 32 bolt.

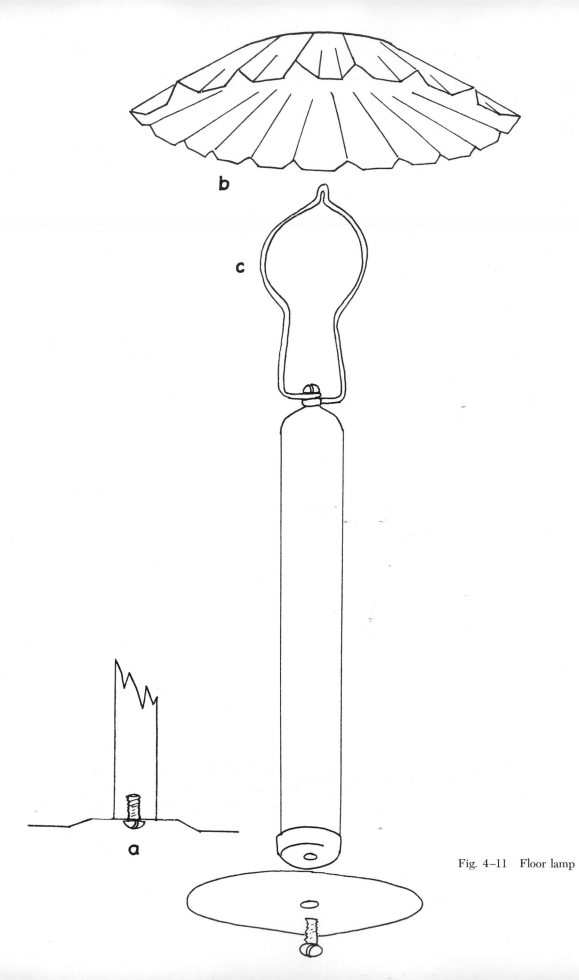

b

c

a

Fig. 4–11   Floor lamp

Pull the end off the cigar tube and punch a hole in it. Bolt the disk to the base of the tube.

Take the lid or bottom from your 1-gallon round can and trim just enough off the edges of this 6″-diameter disk to remove the roughness. This disk will form the lampshade (Fig. 4–11b). Find the center of disk. Place one point of your compass in center, place other point of compass 1″ out from center, and draw a circle. Punch a hole in the center.

Divide disk into sixteen equal pie-shaped sections (just mark—*do not* cut). Place needle-nose pliers on the righthand side of each line, one at a time, and give a slight clockwise twist. Place pliers on lefthand side of each line and give a slight counterclockwise twist. Be sure the tip of your pliers rests on, but does not cross over, the circle you drew 1″ out from center, as you work up the above design.

Using the piece of 16-gauge wire, form the harp (keyhole-shaped gizmo) that holds the lampshade to the upper end of the cigar tube (Fig. 4–11c).

Punch a hole in the tip of the cigar tube just large enough to accommodate a 6 x 32 bolt. Bolt the wire harp to the tip of tube, placing the bolt down through the wires and then through the tube. I put the nut on the end of a pencil-shaped magnet and inserted it down into the tube onto the end of the bolt. Stand tube back onto its base and put lampshade on top. Paint.

LOVE SEAT

To make the love seat shown in Fig. 4–12, you'll need one 1-quart paint thinner or charcoal starter can, approximately $2^3/8″$ x $4^5/8″$.

Remove top and top rim. Remove side seam. Shorten can to about 5″ from bottom. Cut a slit from the outer edge down to the bottom in each of the four corners. Cut three $1/8″$-wide strips down on each side of the corner cuts. (These

Fig. 4–12   Love seat, coffee table, and chaise longue

six strips in each corner will form the legs.) Hold the can bottom-side-up. Twist the three strips to the left of each corner cut one-quarter turn counterclockwise. Twist the three to the right of each corner one-quarter turn clockwise. Clamp these six together about 3/4″ below rim of can. Leave these four legs for the time being.

Divide the two end panels into four equal sections. Divide each of these sections into three equal strips. Shorten one of the side panels to about 2″ from rim. This will be the front of the love seat.

Divide the two side panels into eight equal sections. Divide each of these sections into three equal strips.

On this short side or front of the love seat, remove every other strip (1, 3, 5, and so on). Twist the remaining strips one-quarter turn clockwise and curl them down toward the left into the empty spaces (Fig. 4–13).

Fig. 4–13   Design for front of love seat

On either end, number the strips between the legs 1 through 12. Remove strips 1, 5, and 9. Bend strips 3, 7, and 11 straight up. Shorten the remaining strips—2, 4, 6, 8, 10, and 12—to about 2″ from bottom of can. Give them all one-quarter twist clockwise and curl them down into the empty spaces (Fig. 4–14a). Work up the opposite end in the same manner.

Along the back, number the strips between the legs 1 through 24. Bend the first six straight up. Bend the last six straight up. Bend strips 9, 10, 14, 15, and 16 straight up (Fig. 4–14b).

Remove strip 12. Shorten strips 7, 8, 11, 13, 17, and 18 to about 2″ and curl them down into the empty spaces.

Clamp strips 2, 3, 4, 5, and 6 together 1/2″ from bottom of can. Shorten strips 2 and 6 to about 2″ from clamp and curl them down over clamp. Work up the other two groups of five on the back in a like manner. Finish the back and arms as shown in Fig. 4–14.

Now finish the four legs. Check to make sure the clamps did not move (they should be 3/4″ below rim of can). Shorten strips 1 and 6 to about 2″ from clamp and curl them outward and up over the clamp. Shorten strips 2 and 5 to about 2″ from clamp and curl them inward to a spot just below the clamp. Place a clamp on strips 3 and 4 about 3/4″ below first clamp. Shorten these strips to about 2″ from clamp and curl them down and outward to the clamp (Fig. 4–14c). Paint and add a cushion.

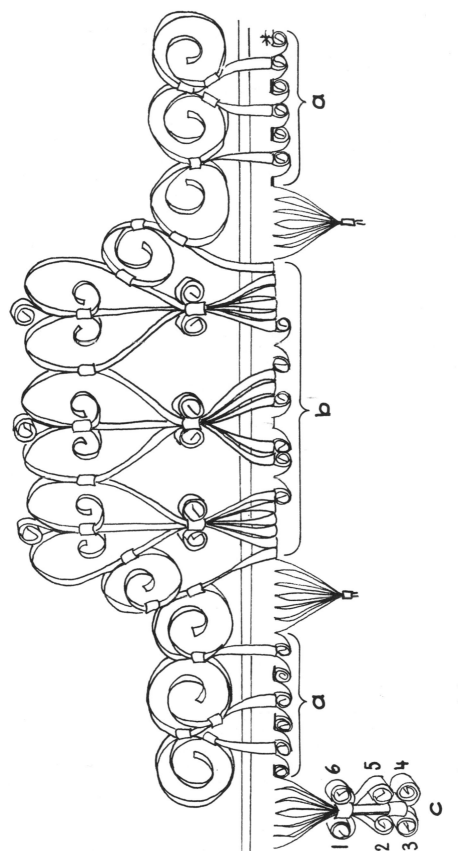

Fig. 4–14 Working pattern for love seat

CHAISE LONGUE

To make this project, shown in Fig. 4–12, you'll need one 1-quart paint thinner or charcoal starter can, approximately $2^3/_8''$ x $4^5/_8''$.

Remove top and top rim; remove side seam. Shorten can to about 5" from bottom. Cut a slit from the outer edge down to the bottom in each of the four corners.

Cut three $^1/_8''$-wide strips down on each side of the corner cuts: These six strips in each corner will form the legs.

Hold the can bottom-side-up. Twist the three strips on the left one-quarter turn counterclockwise. Twist the three on the right one-quarter turn clockwise. Clamp these six together $^3/_4''$ below rim of can. Leave these four legs for the time being.

Shorten one of the end panels to about 2": This will be the foot of the longue. Leave the other end, or the head, full length.

Divide these end panels into two equal sections. Divide each section into six equal strips. Divide the side panels into four equal sections. Divide each section into eight equal strips.

Go to the end of the longue with the short 2" strips. Number the strips 1 through 12, working from left to right. Remove strips 1, 2, 4, and 5. Twist strips 3 and 6 one-quarter turn clockwise. Remove strips 8, 9, 11, and 12. Twist strips 7 and 10 one-quarter turn counterclockwise. Clamp strips 6 and 7 together $^5/_8''$ from bottom of can. Bend strip 6 to the left 90°. Bend strip 7 to the right 90° just below the clamp. Bend strip 3 to the left 90°. Clamp strip 6 to strip 3 just beyond the bend (Fig. 4–15a). Bend strip 10 to the right 90°. Clamp strip 7 to strip 10 just beyond the bend. Strip 3 will work into the leg on the left. Strip 10 will work into the leg to the right.

Turn the chaise longue one-quarter turn toward the right and number the strips 1 through 32. Work from left to right, or from the head to the foot.

Bend strips 1, 5, 9, and 13 straight up. Remove strips 2, 4, 6, 8, 10, 12, 14, 16, 17, 19, 20, 21, 23, 24, 25, 27, 28, 29, 31, and 32 (Fig. 4–15b).

Twist strips 3, 7, 11, and 15 one-quarter turn clockwise. Bend them 90° to the left, $^5/_8''$ from bottom of can. Twist strips 18, 22, 26, and 30 one-quarter turn counterclockwise: $^5/_8''$ from bottom of can, bend them 90° to the right. Clamp them as shown in Fig. 4–15b.

Work up the opposite side in a like manner, only work from right to left, or head to foot.

Number the strips at the head of the longue 1 through 12. Bend strips 1, 3, 5, 8, 10, and 12 straight up (Fig. 4–15c).

Remove strips 4 and 9. Twist strips 2 and 6 one-quarter turn clockwise. Twist strips 7 and 11 one-quarter turn counterclockwise.

Clamp strips 6 and 7 together $^5/_8''$ from bottom of can. Bend strip 6 to the left 90° and strip 7 to the right 90°, just below the clamp. Finish up just the same as other end.

Clamp strips 1, 3, 5, 8, 10, and 12 together $^3/_4''$ from bottom of can. Work up the rest of the head and two sides as shown in Fig. 4–15.

Work up the four legs the same as the legs for the love seat. Paint. Make a cushion to fit.

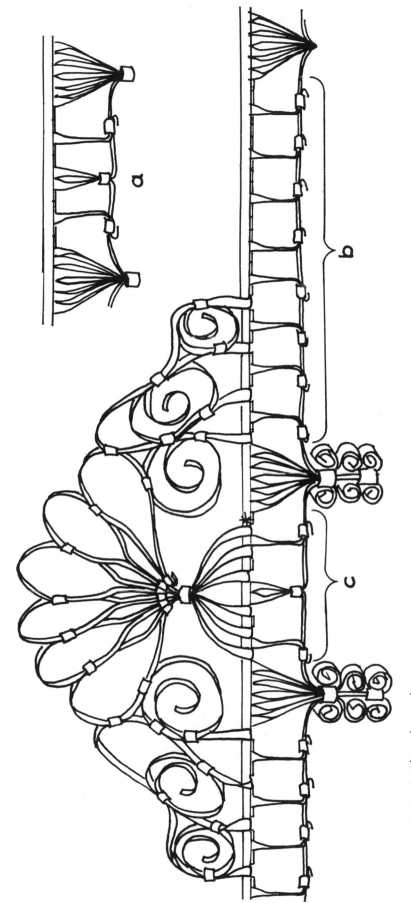

Fig. 4-15 Chaise longue layout

TEA CART

To make the tea cart in Fig. 4–16, you'll need two 3³/₄″ x 1⁷/₈″ oblong cans (Spam or canned beef cans fit these measurements, but try to get a can that does not open around the middle. You will need strips between 2¹/₂″ and 3″ long. If the cans that open through the center are all you can find, it will be necessary to add strips at the corners to keep the top of the tea cart the proper distance from the bottom tray); two 2⁵/₈″-diameter soup cans (for wheels).

One of your oblong cans will be used to make the top tray. Shorten can to about 2″. Locate each of the four corners and cut a slit from the edge down to the rim.

Divide the two end panels into two equal sections each. Divide each section into eight equal strips. Divide each side panel into four equal sections. Divide each section into eight equal strips. You should now have a total of ninety-six strips.

Starting at the space where the seam was removed and working toward your right, bend all the odd-numbered strips straight up 180°. Twist them all one-quarter turn clockwise and curl them down tight toward the right. (Twist them down until they rest on the rim of the can.)

Twist all the even-numbered strips one-quarter turn clockwise and curl them up tight toward the right into the empty spaces.

From the bottom of this tray—or what was originally the *inside* of the can—punch four slots (one in each corner) with a small screwdriver or some other such object (Fig. 4–17a). The four corner posts from the bottom tray will eventually be inserted into these slots to connect the two together. Lay this part aside for the time being.

Fig. 4–16   Ice cream parlor chair, glass-top table, and tea cart

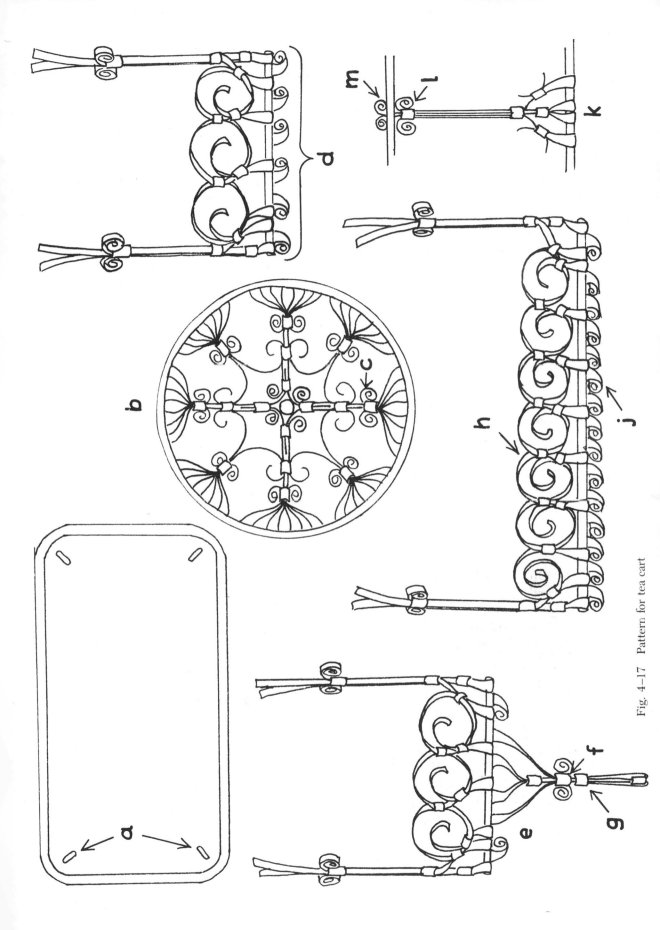

Fig. 4-17 Pattern for tea cart

To make the wheels, remove the tops, top rims, and side seams from two $2^5/8''$ soup cans. Remove the bottoms.

Working with one can at a time, shorten can to about 2" from bottom. Divide can into eight equal sections. Divide each section into eight equal strips. Hold the can right-side-up and number the strips 1 through 64, starting at the space where the seam was removed and working left to right (Fig. 4–17b).

Remove all the even-numbered strips. Twist strips 1, 3, 9, and 11 one-quarter turn counterclockwise. Twist strips 5, 7, 13, and 15 one-quarter turn clockwise.

Clamp strips 1, 3, 5, and 7 together about $3/8''$ from bottom of can. Curl strips 1 and 7 down tight over the clamps (Fig. 4–17c). Clamp strips 9, 11, 13, and 15 together. Curl strips 9 and 15 down tight over the clamps. Put a clamp on strips 11 and 13 about $1^1/8''$ from bottom of can. (This clamp will be temporary, to be replaced later on.)

Now move to the next group of eight strips, on the right, and work them up in the same manner.

Work on around the can, in groups of eight, until all thirty-two strips are worked up into eight groups of four each. Bend all eight groups 90° toward center of can.

Bend strips 13, 29, 45, and 61 to the right 90°. Remove the temporary clamp and slip strip 13 in between strips 27 and 29. Now replace the clamp $1^1/8''$ from bottom of can. Curl strip 11 down over the clamp. Move to the right and slip strip 29 in between strips 43 and 45. Replace clamp and curl strip 43 down over clamp. Move on around to the right and finish the other two groups in a like manner.

Curl strips 3, 19, 35, and 51 in large sweeping curls to the left. Curl strips 5, 11, 37, and 53 in large curls to the right. Clamp them in place as shown in Fig. 4–17b.

For an axle, I used a $2^1/2''$ piece of $1/8''$ aluminum clothesline wire. Force the wire through the center of each wheel and peen the ends down a little with a hammer.

Lay the wheels aside for the time being.

The second oblong can will now be used to make the bottom tray. Remove top and top rim. Remove side seam. Locate each of the four corners and cut a slit from top edge of can down to the bottom rim. Divide the panels on each end of can into two equal sections. Divide each section into eight equal strips.

Hold the can bottom-side-up and number each strip 1 through 16. Work from left to right, starting at end of can where seam was removed.

Remove strips 6, 9, and 13. Bend strips 1, 3, 4, 8, 11, 14, and 16 straight up 180°. Shorten strips 2, 5, 7, 10, 12, and 15 to about 1". Twist strips 2 and 5 one-quarter turn clockwise and curl them down tightly toward the left. Twist strips 7, 9, 12, and 15 one-quarter turn counterclockwise and curl them down to the right (Fig. 4–17d).

Strips 1, 3, 14, and 16 will be part of the corner posts. Leave them for a moment.

Twist strips 4, 8, and 11 one-quarter turn counterclockwise and then curl them in large sweeping curls reaching about ⁵/₈″ above bottom of can. Clamp them as shown in Fig. 4–17d.

Now go to the opposite end of can and number the strips 1 through 16. Bend strips 1, 3, 4, 8, 11, 14, and 16 straight up 180°. Remove strips 6, 9, and 13 (Fig. 4–17e).

Shorten strip 2 to about 1″. Twist it one-quarter turn clockwise and curl it down tight to the right. Shorten strip 15 to about 1″. Twist it one-quarter turn counterclockwise and curl it down tight to the left.

Strips 5, 7, 10, and 12 will form the end leg. Twist strips 5 and 7 one-quarter turn counterclockwise. Twist strips 10 and 12 one-quarter turn clockwise.

Clamp strips 7 and 10 together ³/₈″ below bottom of can. Clamp strips 5 and 12 to the first pair just below the first clamp (Fig. 4–17f). Curl strips 5 and 12 up over the last clamp. Place another clamp on strips 7 and 10 an inch below bottom of can. Bend the long ends left below this clamp back 180° onto themselves and clamp them just below the curls (Fig. 4–17g).

Twist strips 4, 8, and 11 one-quarter turn counterclockwise. Curl them the same as curls on other end.

Divide each side panel into four equal sections. Divide each section into eight equal strips. Each of the two sides will work up in the same manner. Hold the can with the leg end to your left.

Number the strips 1 through 32. Bend strips 1, 3, 5, 9, 13, 17, 21, 25, 29, 30, and 32 straight up 180°. Strips 1, 3, 30, and 32 will be part of the corner posts. Leave them for a moment.

Twist strips 5, 9, 13, 17, 21, 25, and 29 one quarter turn clockwise; then make large curls to match the curls on the ends (Fig. 4–17h).

Remove strips 7, 11, 15, 19, 23, and 27. Shorten the remaining strips to about 1″ and twist them one-quarter turn counterclockwise and curl them down tight to the left (Fig. 4–17j). Do not curl strip 24 on the first side and strip 10 on the opposite side down tightly until the axle for the wheels has been inserted.

Disregard all the previous numbers and renumber the strips in any corner, 1 through 4, left to right. Twist strips 1 and 2 one-quarter turn counterclockwise. Twist strips 3 and 4 one-quarter turn clockwise. Clamp strips 2 and 3 together as close to the bottom rim as you can get (Fig. 4–17k). Clamp strips 1 and 4 to strips 2 and 3, ¹/₂″ above bottom of can. Clamp all four strips together again, about 1³/₄″ from bottom of can. Curl strips 1 and 4 down over this clamp (Fig. 4–17l). Work all four corner posts up in the same manner.

Push the four posts through the slits in the top tray. Curl the protruding strips down tight to hold tray in place (Fig. 4–17m).

Insert the axle on the underside of the tea cart and then curl strips 24 and 10 up tightly to hold axle in place.

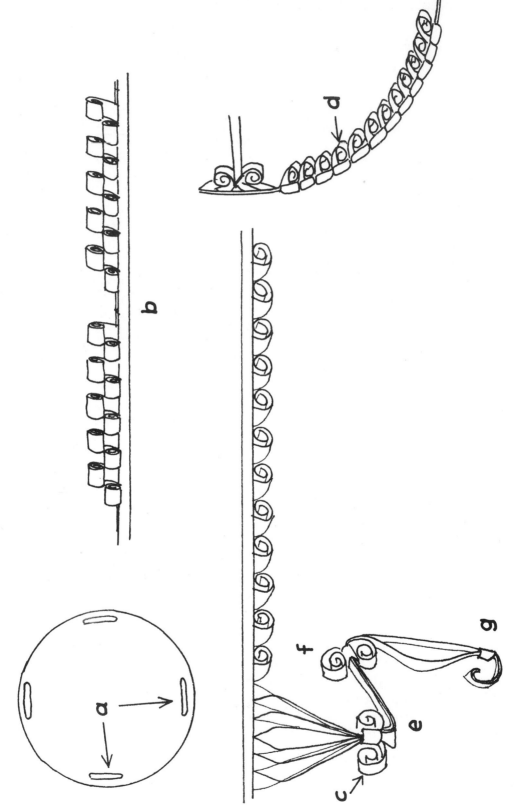

Fig. 4-18 Working pattern for glass-top table

GLASS-TOP TABLE

Materials needed for this project (Fig. 4–16) are one tall juice can, $4^1/4''$ in diameter; one small juice can, $2^1/2''$ in diameter; glass for top of table, flowers.

Remove the top and top rim from the small can. Remove side seam. Shorten can to $1^1/2''$ above bottom. With can opener or other sharp instrument, punch four evenly spaced $1/4''$ slits in bottom of can next to rim (Fig. 4–18a). The legs of the big can will fit through these slits.

Cut sides of can into $1/8''$ strips; total number is of no consequence. Curl these strips down and try to make your finished product resemble a basket (Fig. 4–18b). Remove two strips at each slit to give the table legs room to pass through. Set aside.

Now remove top and top rim from the large can. Remove bottom, but *do not* remove bottom rim. Remove seam. Divide and cut into sixteen equal sections (*see* Ch. 1).

Holding can with bottom up, mark every fourth section. These sections will be left full length for the time being. Cut each of the remaining twelve sections (four groups of three each) down to a length of $1^1/2''$ (Fig. 4–19a).

Divide each of the four long sections into six equal strips (these will be the legs of the table). On each group of six, twist three one-quarter turn clockwise and three one-quarter turn counterclockwise. Clamp all six together $1^1/4''$ from base of can. Cut about $3''$ off the two outside strips and curl them up over the clamp (Fig. 4–18c). Leave these for now and go on to the twelve short sections.

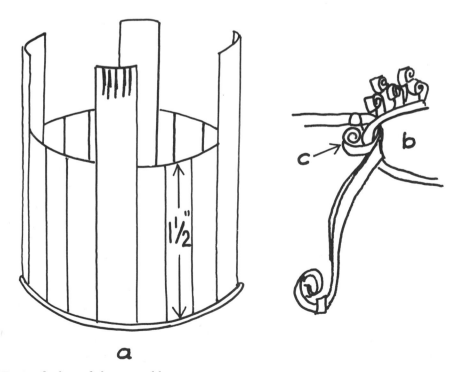

Fig. 4–19   Design for legs of glass-top table

Divide each of these sections into eight strips. Bend every other strip in toward the center of can and twist one-quarter turn clockwise. Using a small curling tool, curl each of these strips in tight (Fig. 4–18d). These curls will form the shelf or ridge on which the glass top will rest.

Now go back to the four legs: $1^1/4''$ down the leg from the first clamp, put a second clamp. Using a heavy pair of pliers, give one-quarter twist to the legs just below the first clamp (Fig. 4–18e). Cut about 2" off each of the inside strips and curl them up over the second clamp (Fig. 4–18f). This will leave three strips in each leg. Bend these strips straight down.

Now take the small can that you prepared for the basket, and put the four legs from the large can through the slots (Fig. 4–19b). Cut about 2" off each of the outside strips and curl them up tight. This will secure the flower basket to the table (Fig. 4–19c).

You will now have two strips remaining for each leg. Clamp these two together about $1^1/2''$ below bottom of basket. Snip about 1" off the ends and make a loose curl for the leg bottom (Fig. 4–18g). Paint desired color. Add glass top: Your hardware store will cut one to size for you from scrap glass. If you prefer, you can cut one from the plastic top of a coffee can or from clear acrylic.

ICE CREAM PARLOR CHAIR

To make this project (Fig. 4–16), use one $2^1/8''$-diameter can ($5^1/2$-ounce juice can).

Remove top, top rim, and side seam. Divide into eight sections. Divide each section into eight equal strips (see Ch. 1).

Hold the can bottom-side-up and number each strip from 1 through 64, starting to the left of the space where seam was removed (asterisk in Fig. 4–20).

Remove strips 2, 4, 6, 8, and 10. Shorten strips 1, 3, 5, 7, 9, and 11 to about $1^1/2''$ from bottom. Twist them one-quarter turn counterclockwise. Curl them up tight toward your right and into the empty spaces (Fig. 4–20a). Save the strips you remove; use them to clamp your work and to add sections to chair back.

Twist strips 12, 13, and 14 one-quarter turn clockwise. Skip over strip 15 for the moment. Twist strips 16, 17, and 18 one-quarter turn counterclockwise (Fig. 4–20b). Strips 12, 13, 14, 16, 17, and 18 make up one of the front legs. Clamp them together 1" below bottom of can. Shape top of leg as shown in Fig. 4–20c. Cut off the two outer strips (12 and 18) about $1/8''$ below clamp. Bend them up over the clamp (Fig. 4–20d). This leaves four strips in the leg. Put a clamp on these four $1^1/4''$ below the top clamp. Cut off the two outer strips (13 and 17) about $1/8''$ below clamp. Bend them up over the clamp. This should leave two strips (14 and 16): Curl them in tightly (Fig. 4–20e).

Remove strips 19, 21, and 23. Shorten strips 20, 22, and 24 to about $1^1/2''$ from bottom. Twist them one-quarter turn counterclockwise. Curl them up tight toward your right into the empty spaces (Fig. 4–20f).

Twist strips 25, 26, and 27 one-quarter turn clockwise. Skip over 28 for the moment. Twist strips 29, 30, and 31 one-quarter turn counterclockwise. These

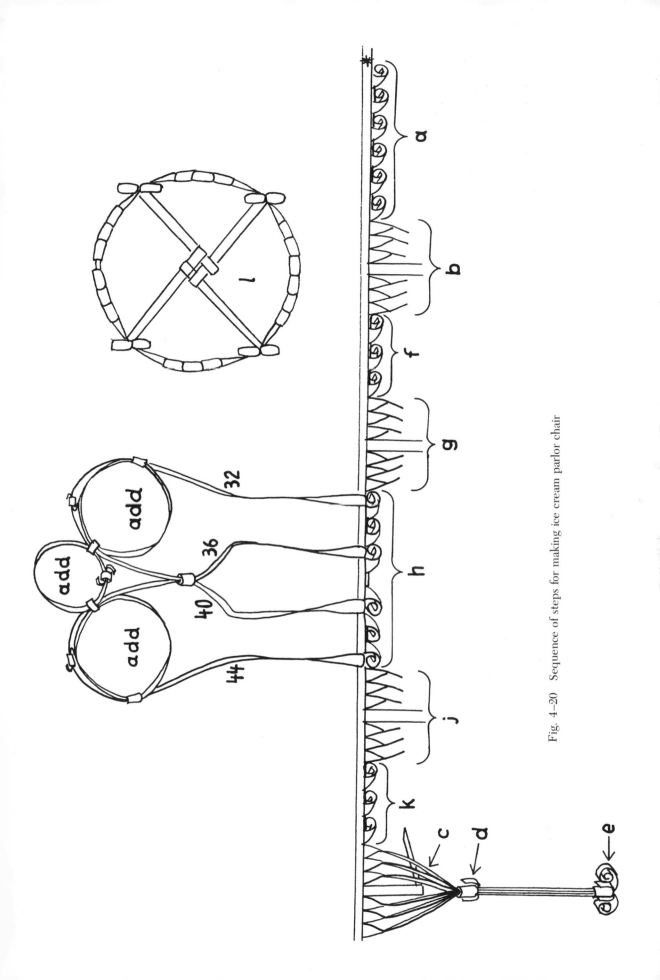

Fig. 4-20 Sequence of steps for making ice cream parlor chair

two groups of three form one of the back legs. Work it up as you did the front one (Fig. 4–20g).

Bend strips 32, 36, 40, and 44 straight up (these will form the back of the chair). Leave them for a moment.

Remove strips 34, 38, and 42. Shorten strips 33, 35, 37, 39, 41, and 43 to about 1¹/₂″ from bottom and curl them up tight (Fig. 4–20h).

Twist strips 45, 46, and 47 one-quarter turn clockwise. Skip over strip 48 for the moment. Twist strips 49, 50, and 51 one-quarter turn counterclockwise. These two groups of three form the other back leg. Work it up as you did the others (Fig. 4–20j).

Remove strips 52, 54, and 56. Shorten strips 53, 55, and 57 to about 1¹/₂″ from bottom of can, twist them one-quarter turn counterclockwise and curl them up tight to your right into the empty spaces (Fig. 4–20k).

Twist strips 58, 59, and 60 one-quarter turn clockwise. Skip over strip 61 for the moment. Twist strips 62, 63, and 64 one-quarter turn counterclockwise. These two groups of three form the other front leg. Form them as you did the other three legs (Fig. 4–20c).

About ³/₄″ down from bottom of can, bend strips 15, 28, 48, and 61 at a sharp right angle toward center of can. Cross these strips and fasten them (Fig. 4–20l).

Twist strips 32 and 36 one-quarter turn clockwise. Twist strips 40 and 44 one-quarter turn counterclockwise about 1¹/₂″ from bottom of can. Clamp strips 36 and 40 together. Then clamp strips 32 and 36 together and strips 40 and 44 together. Using the long strips that were removed, make three rings: two about ⁷/₈″ in diameter; one about ⁵/₈″ in diameter. Fasten these to back of chair and shape chair (Fig. 4–20 add).

STRAIGHT-BACK CHAIR

One soup can, 2⁵/₈″ in diameter, is the only material required to make this chair.

Remove top and top rim. Remove seam. Divide can into eight equal sections (*see* Ch. 1). Divide each section into eight equal strips. Hold the can bottom-side-up and number each strip from 1 through 64; work toward your left, starting to the left of space where seam was removed.

Remove strips 2, 4, 6, 8, and 10. Do not throw these strips away: You will need them for clamping and to complete the back design. Shorten strips 1, 3, 5, 7, 9, and 11 to about 2″. Twist them one-quarter turn counterclockwise and curl them into the spaces where strips were removed (Fig. 4–21a).

Twist strips 12, 13, and 14 one-quarter turn clockwise. Twist strips 15, 16, 17, and 18 one-quarter turn counterclockwise. Clamp these seven together about 1³/₈″ from bottom of can. This is one of the front legs (Fig. 4–21b).

Remove strips 20 and 24. Bend strips 22 and 26 straight up. Shorten strips 19, 21, 23, and 25 to about 2″; twist them one-quarter turn clockwise and curl them into empty spaces (Fig. 4–21c).

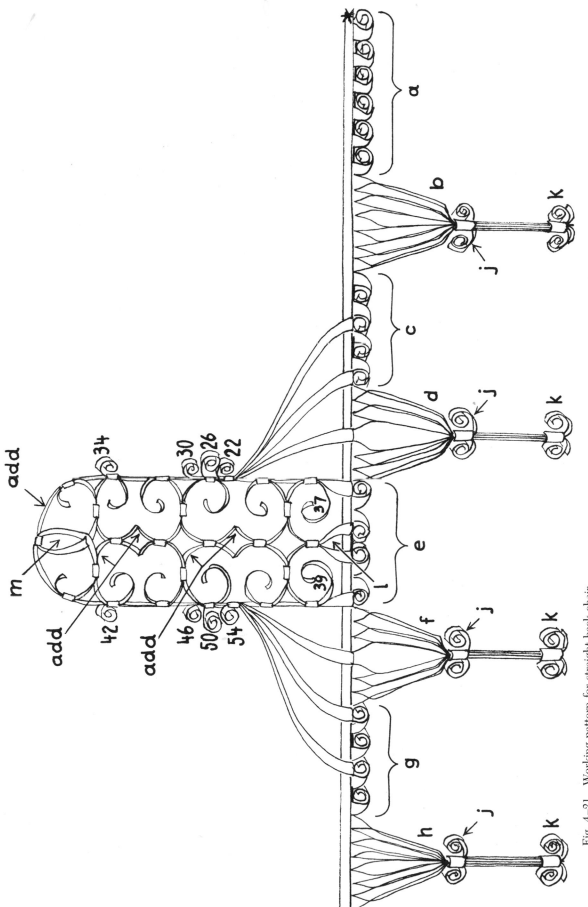

Fig. 4-21 Working pattern for straight-back chair

Twist strips 27, 28, and 29 one-quarter turn clockwise. Bend strip 30 up. Twist strips 31, 32, and 33 one-quarter turn counterclockwise. Clamp this group of six together 1³/8″ from bottom of can. This is one of the back legs (Fig. 4–21d).

Bend strips 34, 37, 39, and 42 straight up. Remove strip 38. Shorten strips 35, 36, 40, and 41 to about 2″ and curl them into the empty spaces (Fig. 4–21e).

Twist strips 43, 44, and 45 one-quarter turn clockwise. Bend strip 46 straight up. Twist strips 47, 48, and 49 one-quarter turn counterclockwise. Clamp this group of six together 1³/8″ from bottom of can. This forms the other back leg (Fig. 4–21f).

Bend strips 50 and 54 up. Remove strips 52 and 56. Shorten strips 51, 53, 55, and 57 to about 2″ and twist them one-quarter turn counterclockwise and curl them into the empty spaces (Fig. 4–21g).

Twist strips 58, 59, 60, and 61 one-quarter turn clockwise. Twist strips 62, 63, and 64 one-quarter turn counterclockwise. Clamp these seven together about 1³/8″ from bottom of can. This forms the other front leg (Fig. 4–21h).

Now, go back and finish the four legs. Check to make sure the clamps did not move. (They should each be 1³/8″ from bottom of can.) Shorten the outside strips on each leg—12, 18, 27, 33, 43, 49, 58, and 64—to about 2″ from clamp. Curl them up over the clamps (Fig. 4–21j). Put another clamp 2¹/2″ from bottom of can on each of the four legs. Curl the outside strips—13, 17, 28, 32, 44, 48, 59, and 63—up over the clamps (Fig. 4–21k). Cut off the remaining strips—14, 15, 16, 29, 31, 45, 47, 60, 61, and 62.

Now, finish the chair back. Clamp strips 37 and 39 together ¹/4″ above bottom of can. Cut about 1″ off the end of each of these two strips. Make large sweeping curls (Fig. 4–21l).

Clamp strips 22, 26, 30, and 34 together about 1¹/4″ above bottom of can. Clamp strips 42, 46, 50, and 54 together as shown in diagram. Curl strips 22 and 54 down over clamps.

Take four of the strips that were removed from can and make a design for the back, as shown. Using another strip that was removed, add it on to extend the chair back about ³/4″. Now add one more strip of trim (Fig. 4–21m).

Make a Styrofoam or foam rubber pad to fit chair bottom and cover with suitable material (see Fig. 2–18).

BED

To make the bed shown in the color section, Fig. 6, you'll need one 2-quart rectangular charcoal starter fluid can, approximately 5¹/8″ x 3¹/2″ at the base.

Remove top and top rim. Locate the center of each of the four corners and cut a slit from top edge down to the base of the can. Then cut five strips (each approximately ¹/8″ wide to each side of the four corner cuts. You will now have four groups with ten strips each (one group in each of the four corners). Holding the can bottom-side-up, number the strips in each group 1 through 10 (left to right). Bend strips 2 and 4 up and twist them one-quarter turn clockwise. Bend strips 7

Fig. 4–22   Dressing table, bed, and bedside table

and 9 up and twist them one-quarter turn counterclockwise. Clamp these four together about ³/₄″ above base of can. Twist strips 1, 3, and 5 one-quarter turn counterclockwise. Twist strips 6, 8, and 10 one-quarter turn clockwise. Clamp these six strips together about ³/₄″ below base of can. These groups of six will form the four legs of the bed (Fig. 4–23a).

Shorten the large panels on each side of the can to about 3″. Divide each of these two larger panels into four equal sections. Divide each of these four sections into eight equal strips (this will give you a total of thirty-two strips on each side). Number them 1 through 32, from left to right.

Remove strips 1, 3, 5, 7, 9, 11, 13, and 15 (watch it) 18, 20, 22, 24, 26, 28, 30, and 32. Twist strips 2, 4, 6, 8, 10, 12, 14, and 16 one-quarter turn clockwise. Twist strips 17, 19, 21, 23, 25, 27, 29, and 31 one-quarter turn counterclockwise. At a point ¹/₂″ below base of can, bend strips 2, 4, 6, 8, 10, 12, 14, and 16 at a sharp 90° angle to the left. Clamp them as shown in Fig. 4–23b. At a point ¹/₂″ below base of can, bend strips 17, 19, 21, 23, 25, 27, 29, and 31 at a sharp 90° angle to the right. Clamp them as shown in Fig. 4–23b.

To form the foot of the bed, shorten the panel between the legs to approximately 4″. Divide this panel in half. Then divide each half into eight equal strips. This will give you sixteen strips. Number them 1 through 16, left to right (Fig. 4–23c).

Bend strips 1, 3, 5, 7, 10, 12, 14, and 16 up. Twist strips 1 and 3 one-quarter turn clockwise. Twist strips 5 and 7 one-quarter turn counterclockwise. Clamp these four strips together ³/₄″ above base of can (Fig. 4–23d).

Twist strips 10 and 12 one-quarter turn clockwise. Twist strips 14 and 16 one-quarter counterclockwise. Clamp these four strips together ³/₄″ above base of can (Fig. 4–23e).

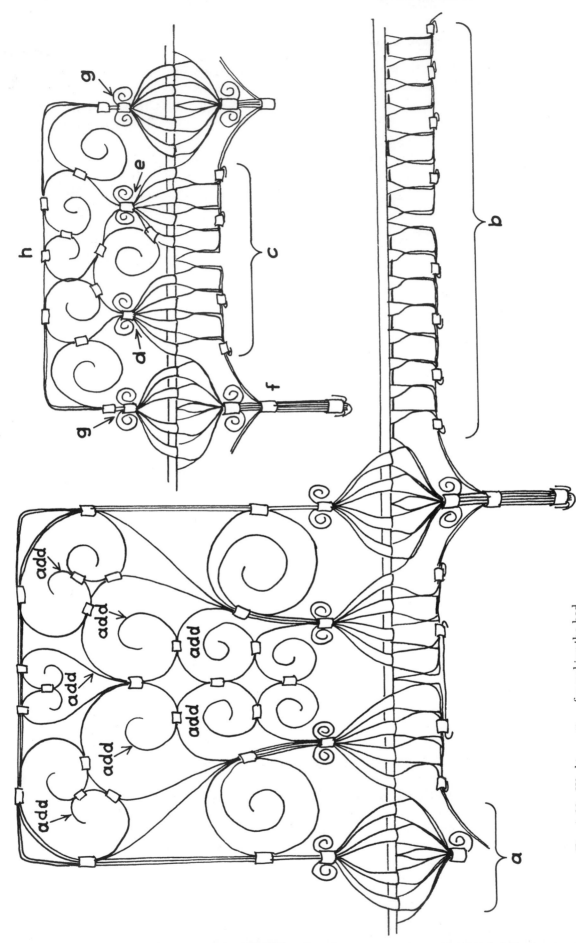

Fig. 4–23 Working pattern for making the bed

Twist strips 2, 4, 6, and 8 one-quarter turn clockwise at a point $1/2''$ below base of can. Bend these four strips at a sharp 90° angle to the left. Twist strips 10, 12, 14, and 16 one-quarter turn counterclockwise. At a point $1/2''$ below base of can, bend these four strips at a sharp 90° angle to the right. Clamp them and form the legs (Fig. 4–23f).

At the foot of the bed you should now have four groups of four strips each. Forget the numbers previously assigned to these strips and renumber them, 1 through 16. Snip off strips 1, 4, 5, 8, 9, 12, 13, and 16 to about $1 1/2''$ above the clamp. Curl them down tight over the clamp (Fig. 4–23g). Now form the remaining eight strips (Fig. 4–23h), or make up a design of your own.

To form the head of the bed, leave the panel between the two legs at its full length. Divide it in half, then divide each half into eight equal strips. Refer back to the paragraphs on how to form the foot of the bed and work the head up in the same manner.

Look at the diagram very carefully, as several of the curlicues in the headboard are extra strips that have been added.

Paint; add a foam rubber mattress if desired.

NIGHTSTAND

Materials needed to make the nightstand shown in color in Fig. 6 are: one $1 3/4''$ x $1 3/4''$ square can (Hickory Farms tea); one $3''$ bolt; one top from a $2 5/8''$ can (for lampshade); one large drinking straw.

I assume you are using a tea can; the reason I assume this is that I searched for a small square can for weeks before I found this $1 3/4''$-square can. You are really lucky—I did all the hunting for you.

The can I used had no top rim, just a bead or a ridge around it about $1/4''$ from the top. Do not remove this ridge: Cut through it as you make your strips. You will need the full $2 3/8''$ length.

Remove the side seam. Removal of the seam will leave about a $3/16''$ gap. Go to each of the three remaining corners and remove a strip this same width.

Divide each of the four side panels in half. Divide each of these halves into seven equal strips. Each of the four sides will be worked up in the same manner. Hold can bottom-side-up! Number the strips on any one of the sides 1 through 14.

Twist strips 1, 2, and 3 one-quarter turn clockwise (Fig. 4–24a). Remove strips 4 and 6. Shorten strips 5 and 7 to about $1''$ from bottom of can, twist them one-quarter turn counterclockwise, and curl them down to the left. Remove strips 9 and 11. Shorten strips 8 and 10 to about $1''$, twist them one-quarter turn clockwise, and curl them down to the right (Fig. 4–24b). Twist strips 12, 13, and 14 one-quarter turn counterclockwise.

In each of the four corners you should now have six long strips. Clamp each of these six together about $3/4''$ from bottom of can. These will form the legs of your nightstand (Fig. 4–24c).

Shorten the two outside strips of each group to about $1''$ from clamp and

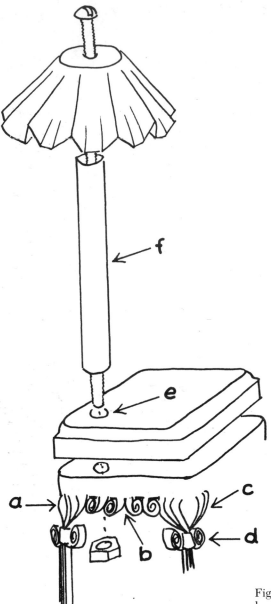

Fig. 4–24 Making a bedside table with built-in lamp

then curl them up over the clamp (Fig. 4–24d). Place a clamp on the remaining four strips 2″ below bottom of can. *Do not* curl; just bend the ends back over the clamp.

Glue or solder the top of the can to what is now the top of the nightstand.

If you do not have a soldering iron and intend to glue the two together, I would suggest that you cut several squares of corrugated cardboard and glue several together until they are thick enough to fill the space between the inside of the lid and the top of the table. Then glue the two sections together.

Punch a hole just big enough to take a #6 bolt, all the way through the top of your nightstand. Punch this hole about ¾″ in from a corner (Fig. 4–24e).

Fig. 1 A child's garden of flowers

Fig. 2
Fruit basket

Fig. 3
Easter basket

Fig. 4 Two candlestands, hurricane lamp, bonbon dish, toothpick holder

Fig. 5 Circus wagon

Fig. 6 Bedroom furniture: nightstand, bed, dressing table and bench, hassock

Fig. 7 Baby furniture: small rocker, cradle, highchair, playpen, stroller

Fig. 8 Birdcage flower basket, hurricane lamp, Americana lamp

Fig. 9 Wheelbarrow and flower cart

Fig. 10
Christmas tree
ornaments

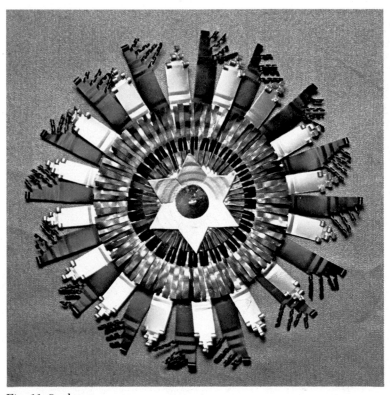

Fig. 11 Sunburst

Fig. 12 Christmas bells

Cut $2^{1}/_{4}''$ off a large plastic drinking straw. This $2^{1}/_{4}''$ tube will fit down over the $3''$ bolt to form the upright of the lamp (Fig. 4–24f).

The $2^{5}/_{8}''$-diameter can top will form the lampshade. Find the center of this disk. Place one point of your compass in the center. Place the other point $^{1}/_{2}''$ out. Draw a circle. Divide the disk into twelve equal pie-shaped sections (*do not cut—just mark*). Place your needle-nose pliers along the righthand side of these lines, one at a time, and give them a slight twist clockwise. Make sure the point of your pliers rests on the circle you drew in the center.

Place your pliers on the left hand side of each line and give a slight counterclockwise twist.

Punch a hole in the center just large enough for the #6 bolt to fit through. Slip the tube over the bolt and fasten to table. Paint.

DRESSING TABLE

To make this project (color section, Fig. 6), you'll need one 1-quart oblong turpentine or charcoal lighter fluid can, $2^{3}/_{8}'' \times 4^{1}/_{2}''$ bottom measurement.

Remove top and top rim. Remove side seam. Put a mark $2^{1}/_{2}''$ from bottom all the way around the can. Locate the four corners, mark, and cut a slit from top edge to rim at bottom of can.

Cut three strips about $^{1}/_{8}''$ wide down each side of the four corner cuts. This will give you six strips in each corner for the legs.

Cut off the remainder of the can on the $2^{1}/_{2}''$ mark (save these pieces). Divide each of the large front and back panels into four equal sections. Divide these eight sections into eight strips each. Divide the two end panels into two sections each. Divide these four sections into *six* strips each.

Hold the can bottom-side-up and go to work on the corner where the seam was removed (marked by an asterisk in Fig. 4–25). If your cuts were the same as mine, you should have three strips between the seam space and the corner cut. Twist these three one-quarter turn clockwise. Take the next three strips on the left and twist them one-quarter turn counterclockwise. Clamp these six together $1''$ below bottom of can (Fig. 4–25a). Shorten the two outside strips to about $1^{1}/_{2}''$ from clamp. Curl them up tight over clamp (Fig. 4–25b). Place a clamp on the remaining four strips $3''$ from bottom of can. Curl the ends up to clamp (Fig. 4–25c). Twist the two strips on the right of the seam space one-quarter turn clockwise, move them toward the left, and clamp them to the leg $^{1}/_{2}''$ from bottom of can. Curl them up to clamp (Fig. 4–25d).

Twist the two strips on the left of the leg one-quarter turn counterclockwise. Move them toward the right and clamp them to the leg $^{1}/_{2}''$ from bottom of can. Curl them up to the clamp (Fig. 4–25e). This completes leg number one. Go to each of the remaining three corners and work up a leg in the same manner.

Come back now to the seam end. If we have been working in harmony, you should have eight strips between the two legs. Remove the center two. Twist the three on the right one-quarter turn clockwise. Twist the three on the left one-quarter turn counterclockwise. Clamp these six together $^{3}/_{4}''$ below bottom of

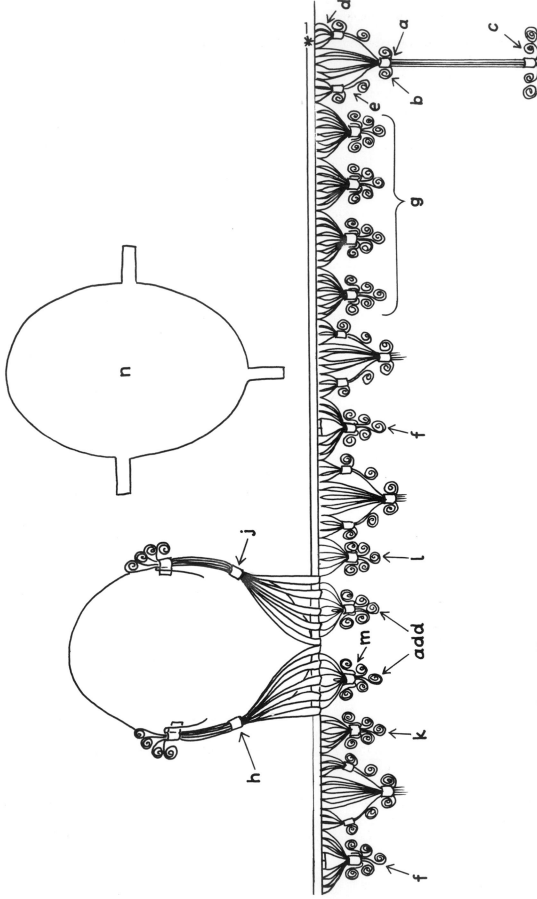

Fig. 4–25 Layout for dressing table

can. Snip one of the outside strips off $1/8''$ from the clamp and bend it down over the clamp. Curl the remaining five down to the clamp (Fig. 4–25f).

Move to the far end now. There should be eight strips here. Work them up in the same manner as the other end.

Choose either side for the front of the dressing table.

There should be twenty-eight strips between the two front legs. Number them 1 through 28. Twist strips 1, 2, 3, 8, 9, 10, 15, 16, 17, 22, 23, and 24 one-quarter turn counterclockwise. Twist strips 4, 5, 6, 7, 11, 12, 13, 14, 18, 19, 20, 21, 25, 26, 27, and 28 one-quarter turn clockwise. Clamp these together in groups of seven (four groups), $3/4''$ from bottom of can (Fig. 4–25g).

Snip off the two outside strips of each group $1/8''$ from the clamp and bend them back tight over the clamp. Curl the remaining five down to the clamp.

As the project is getting to look more like a dressing table than a can, we won't call it a can any longer. Turn dressing table around so you can work up the back. Number the strips between the two back legs 1 through 28.

Bend strips 6, 8, 10, 12, and 14 up 180°; twist them all one-quarter turn counterclockwise. Place a clamp on them about $1''$ above dresser top (Fig. 4–25h). Bend strips 15, 17, 19, 21, and 23 up 180°; twist them all one-quarter turn clockwise. Place a clamp on them $1''$ above dresser top (Fig. 4–25j). These ten strips will hold the mirror in place. You will come back to them later.

Twist strips 1, 2, and 3 one-quarter turn counterclockwise. Twist strips 4 and 5 one-quarter turn clockwise. Clamp these five together $3/4''$ below dresser top (Fig. 4–25k).

Move all the way over to far end; twist and curl strips 24, 25, 26, 27, and 28 up in the same manner (Fig. 4–25l).

Before you proceed, cut two $1/8''$-wide strips, about $2''$ long, from the excess pieces discarded at beginning.

Now finish up the back. Twist strips 7 and 9 one-quarter turn counterclockwise. Twist strips 11 and 13 one-quarter turn clockwise.

Place one of the extra strips between strips 9 and 11 and clamp the five together $3/4''$ below dresser top. Curl down to the clamp (Fig. 4–25m). Finish off the remaining four strips and the extra strip in a like manner.

Cut a $3''$ x $2 1/4''$ oval from the excess tin (this oval will be the mirror). Leave a $1/8''$ tab about $1/2''$ long on each side, as well as one at the bottom (Fig. 4–25n).

Follow the diagram to attach mirror to table. Paint.

DRESSING TABLE BENCH

To make a matching bench for the dressing table (color Fig. 6) you'll need one oblong Band-Aid can, $3 1/4''$ x $1 1/2''$ bottom measurement.

Remove top and top rim. Remove side seam. Mark and cut the four corners from top edge down to the bottom rim.

Divide each *end* panel into two equal sections; divide each section into six equal strips. Divide each *side* panel into four equal sections; divide each section into six equal strips.

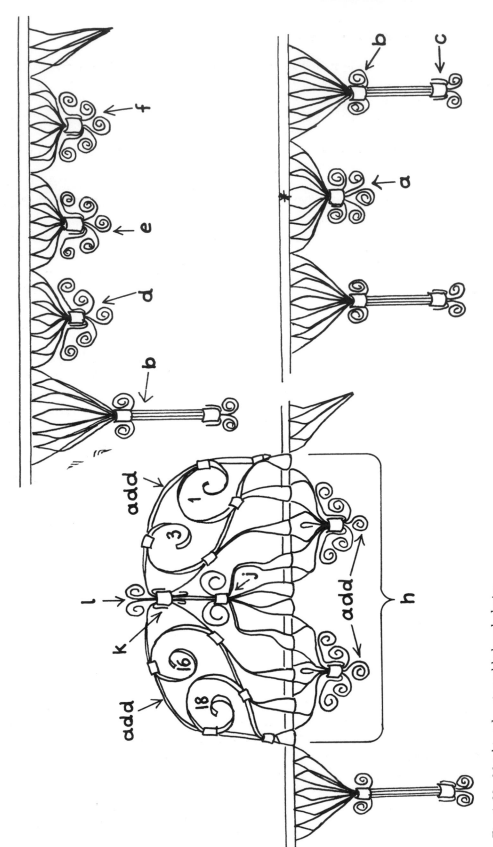

Fig. 4–26 Matching dressing table bench design

Start work on the end of can where seam was removed (asterisk in Fig. 4–26). Work carefully—I notice the tin strips have a tendency to pull out from the rim at this point.

To the left of the space where seam was removed, twist the three strips one-quarter turn counterclockwise. Twist the three strips on the right one-quarter turn clockwise. Clamp these six together $1/2''$ from bottom of can. Shorten one of the outside strips to $1/8''$ above the clamp and then bend it down over the clamp. Shorten the five remaining strips to about $1^1/2''$ from clamp. Curl them down to the clamp (Fig. 4–26a). (This should eliminate the problem of the strips pulling out of the rim.) Now go to the far end and follow the same procedure.

Start to the right of one of these groups and number the strips 1 through 30, working left to right. Twist strips 1, 2, and 3 one-quarter turn counterclockwise. Twist strips 4, 5, and 6 one-quarter turn clockwise. Clamp these six together $1''$ below bottom of can (Fig. 4–26b)—this will be a leg. Curl strips 1 and 6 down over the clamp. Place a second clamp $1''$ below the first one (Fig. 4–26c). Snip off strips 2 and 5, $1/8''$ from clamp, and bend them up over the clamp. Curl strips 3 and 4 down to clamp.

Twist strips 7, 8, and 9 one-quarter turn counterclockwise. Twist strips 10, 11, and 12 one-quarter turn clockwise. Clamp these six together $1/2''$ from bottom of can. Snip strip 7 off $1/8''$ from clamp and bend it up over the clamp. Shorten the remaining five to about $1^1/2''$ from the clamp and curl them down (Fig. 4–26d).

Twist strips 13, 14, 15, 19, 20, and 21 one-quarter turn counterclockwise. Twist strips 16, 17, 18, 22, 23, and 24 one-quarter turn clockwise. Clamp these together in groups of six and curl down in same manner as first group (Fig. 4–26e and f).

This brings us to the second leg: Form it in the same way as the first leg (Fig. 4–26b).

Using six strips at each of the two remaining corners, form the last two legs just as you did the others.

You should now have eighteen strips across the back. Number them 1 through 18, right to left (Fig. 4–26h).

Bend strips 1, 3, 5, 7, and 9 up 180°. Give them all one-quarter twist counterclockwise. Bend strips 10, 12, 14, 16, and 18 up 180°. Twist them one-quarter turn clockwise. Clamp strips 7, 9, 10, and 12 together $3/4''$ above bottom of can (Fig. 4–26j). Shorten strips 7 and 12 to about $1^1/2''$ above clamp. Curl them down over the clamp.

Before you go any further, take time out: You will need four extra $1/8''$-wide strips to complete this project. Get another can or a flat piece of tin you have saved from another can and cut four strips, each about $2^1/2''$ long.

Clamp strips 5 and 14 to strips 9 and 10, $1^1/4''$ above bottom of can. Before you tighten this clamp up, add two of the extra strips: one between strips 5 and 9 and the other between strips 10 and 14. Tighten clamp. Snip strips 5 and 14 off $1/8''$ above clamp and bend them down over the clamp (Fig. 4–26k).

Bend *added* strip 1 in a large circle to your right. Bring the loose end down and clamp it to permanent strip 1 just above the bottom of can.

Bend *added* strip 2 in a large circle to your left and clamp it to permanent strip 18 just above bottom of can. (The original strips were not long enough to make this design. That's the reason for these added strips.)

Curl strips 9 and 10 down tightly on top (Fig. 4–26l). Finish up the back design by fitting strips 1, 3, 16, and 18 into place as shown in diagram.

Twist strips 2, 4, 11, and 13 one-quarter turn counterclockwise. Twist strips 6, 8, 15, and 17 one-quarter turn clockwise.

Clamp strips 2, 4, 6, and 8 together. But before you tighten up the clamp, add another extra strip between 4 and 6. Place clamp about 1/2″ from bottom of can. Curl the five strips down to the clamp.

Clamp strips 11, 13, 15, and 17 together. Add another extra strip between 13 and 15. Clamp and curl, just as before.

Add a cushion and paint.

Chapter  5

Baskets

VIOLET BASKET

Materials needed for this project (Fig. 5–1) are one 2⅝″-diameter can (soup can), and several aluminum beverage cans for violets and leaves.

Remove top and top rim from the 2⅝″-diameter can. Remove side seam. Divide can into eight equal sections (*see* Ch. 1). Divide each section into eight equal strips.

Start at the space where seam was removed and number the strips 1 through 64. Twist each of the *odd-numbered* strips one-quarter turn counterclockwise. Twist each of the *even-numbered* strips one-quarter turn clockwise.

Fig. 5–1   Fruit basket, violet basket, and bell basket

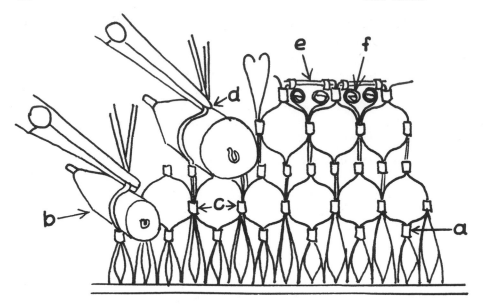

Fig. 5–2   Working pattern for violet basket

Clamp strips 1 and 2 together 1/2″ above bottom of can. Skip strips 3 and 4. Clamp strips 5 and 6 together. Skip strips 7 and 8. Go all the way around can, clamping two and skipping two (Fig. 5–2a).

Form and clamp strips 1 and 2 around a 9/16″ dowel stick—a stick 2″ or 3″ long will work nicely (Fig. 5–2b). Form and clamp the remaining pairs in the same manner.

You should now have sixteen rings and sixteen pairs of unclamped strips. Clamp each ring to its next-door neighbor. Make sure a pair of the unclamped strips are placed between them (Fig. 5–2c).

Clamp together the two strips that come up between each of the 9/16″ rings, about 1/4″ from bottom of can. (These clamps should be even with the clamps on top of the 9/16″ rings.) Form and clamp these same strips around a 3/4″ dowel stick (Fig. 5–2d).

Clamp each of these rings to its next-door neighbor. Make sure that the pair of strips from the 9/16″ ring, just below, is between them. Curl the strips that come from 9/16″ rings down even with the top clamps (Fig. 5–2e). Bend the remaining strips down and clamp them (Fig. 5–2f).

Paint. Add violets and leaves (see Ch. 3 for patterns).

When the violets and leaves are arranged in basket, the top rim is not visible. But if you would like a fancier rim on top, see directions for bell basket.

BELL BASKET

To make the bell basket (Fig. 5–1), you'll need: two cans, 3 1/8″ in diameter and approximately 4 1/8″ tall (one with ridges and one with straight sides); one rim from a 2-pound coffee can; several feet of tin strips, 3/16″ wide; 28-gauge hobby wire.

Start with the ridged can. The more ridges or beads on the can, the more design on the finished basket.

Remove top and top rim. Remove side seam. Divide can into twenty-four equal sections. Each section will be about $7/16''$ wide.

Shorten every second section to about 2″ from bottom of can (Fig. 5–3a). Divide each of these short sections into three strips each (Fig. 5–3b). Twist each of these strips one-quarter turn clockwise and curl them down as shown in Fig. 5–3c.

Make two $1/2''$ slits in the ends of each of the twelve long sections (Fig. 5–3d). Attach the rim from coffee can as shown (Fig. 5–3e). Coil the long $3/16''$-wide strips of tin around a coathanger wire (*see* Fig. 1–20), barber pole fashion. Bind these coils to top of basket with a fine wire (Fig. 5–3f).

The second $3^1/8''$-diameter can will be used to make the base or legs of the basket. Remove top and top rim. Remove side seam. Shorten can to about $3^1/2''$ from bottom. Divide into twelve sections. Divide each section into six strips.

Starting at the space where seam was removed, and working left to right, twist three strips one-quarter turn counterclockwise. Twist the next three one-quarter turn clockwise. Alternating three counterclockwise and three clockwise, go all the way around the can.

Skip strip 1. Gather the next five together and clamp 1″ from bottom of can (Fig. 5–4a). Skip one and gather five all the way around. You should now have twelve groups of five strips each. Curl each of these five down as shown (Fig. 5–4b).

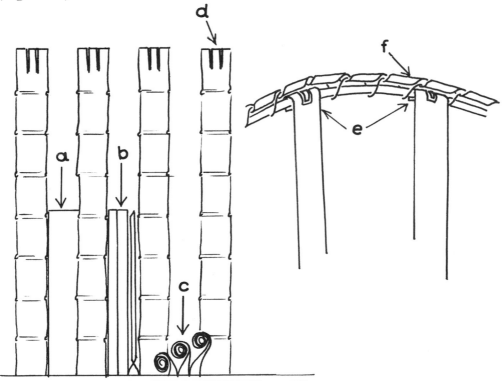

Fig. 5–3   Design for bell basket

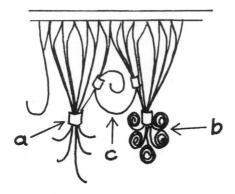

Fig. 5–4   Making legs for bell basket

Form the single strips into large loose curls and bind them on both sides to the groups of five (Fig. 5–4c).

Fasten cans together. Paint.

EASTER BASKET

Materials needed to make this basket (color section, Fig. 3) are one 1-gallon motor oil can; one large ball of twine; about 10' of aluminum clothesline wire, $1/8''$ diameter.

Remove top and top rim from can. Remove side seam. Divide can into an odd number of equal-width strips. The total number of strips is not important, so long as it is an odd number (29, 31, or 33 would be good).

Round off the two corners on the end of each strip. This will help keep down the number of nicks and scratches on your knuckles. Set can on table in an upright position. Bend all the spokes down flat on table. Your can will now resemble a large hub with spokes reaching out about 7'' or 8''. Since the spokes are too close together at the hub to weave, remove a sliver about $1/16''$ wide on one side of each of the spokes. This sliver does not have to be the full length of the spoke, just an inch or two, enough to facilitate the weaving (Fig. 5–5a).

With the twine, weave in and out until you have about 2'' or $2^1/2''$ woven out from bottom of can. Bend all twenty-nine spokes up at a sharp 90° angle. This will form the side of the basket. If you have a large pot, such as a pressure cooker, you will find it helpful to fit the basket down over the pot and then weave the sides right on the pot, using it as a mold. This procedure is not necessary, but it does help to keep the sides straight up. Weave about 2'' or 3'' for the sides. Cut off the spokes, all the same length, about $1/2''$ above the finished weave. Cut a piece of aluminum wire long enough to reach around top rim of basket. With a pair of regular pliers, bend each spoke down and around wire (Fig. 5–5b).

Cut three more pieces of wire, about 26'' long each, for the handle. (The length of these three wires depends on the size of handle desired.) Bend one end of each of these wires around the wire rim. Fasten them to rim about 1'' apart (Fig. 5–6a). Go to the directly opposite side of basket and fasten the other ends of the handle.

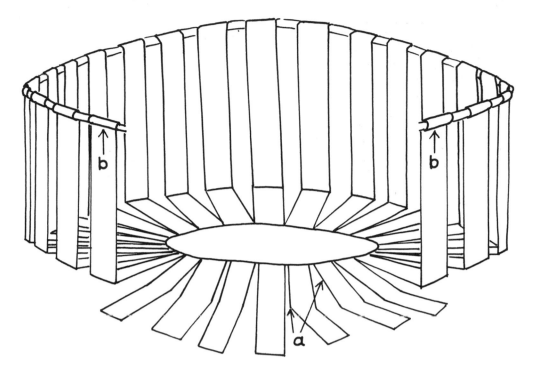

Fig. 5–5   Forming the Easter basket

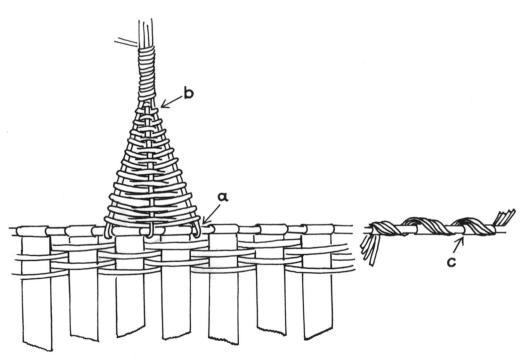

Fig. 5–6   Weaving the handle for Easter basket

Start at the top center of the handle and wrap the three strands of wire together, using the same type string used to weave basket. Wrap for about 6″ or 7″. Then divide the three wires and start an over-and-under weave again (Fig. 5–6b). When you have woven all the way down to the basket rim, tie the string off and go back to the top center of handle; work down the other side in a like manner.

Measure a piece of string about seven or eight times the length around top rim of basket. This should be long enough to wrap around top rim four or five times, in between each spoke and over the wire rim. This will give your basket a finished look (Fig. 5–6c). Spray with a clear lacquer if desired.

FRUIT BASKET

Materials needed to complete this basket, shown in color Fig. 2, are two 4¼″-diameter juice cans and eighteen strips of tin, ³/₁₆″ wide x 8″ long.

The first can will form the main body of the basket. Remove top and top rim. Remove side seam. Divide can into eighteen equal sections. Divide each section into four equal strips; strips will be about ³/₁₆″ wide. Hold the can bottom-side-up. Starting at the space where seam was removed, and working from left to right, twist strips 1 and 2 one-quarter turn counterclockwise. Twist strips 3 and 4 one-quarter turn clockwise. Go all the way around the can, twisting two strips counterclockwise, then two clockwise (Fig. 5–7a).

Clamp strips 1, 2, 3, and 4 together 1″ from bottom of can, then clamp strips 5, 6, 7, and 8 together. Go all the way around the can, clamping groups of four together. When finished, you should have eighteen groups of four each (Fig. 5–7b).

Place a second clamp on each of these groups 3″ from bottom of can (Fig. 5–7c). Tilt these groups out a little from the perpendicular. The top of your finished basket should have about a 7″ diameter.

Hold the can right-side-up. Number the strips in each group 1 through 4. Work left to right. Just above the second clamp (3″ above the bottom of can), bend strips 1 and 2 of *each* group, a sharp 45° to the left. Move ³/₄″ out toward the end of strips and bend back a sharp 45° to the right.

Just above the second clamp, bend strips 3 and 4 of *each* group a sharp 45° to the right. Move out ³/₄″ toward the end of strips and bend back a sharp 45° to the left.

Go all the way around the can, making new groups of four each by clamping strips 1 and 2 of one group to strips 3 and 4 of the adjoining group (Fig. 5–7d). Renumber the strips in these eighteen new groups from 1 to 4. Curl strips 1 and 4 of each group down tight over the clamp (Fig. 5–7e). Bend strip 2 of each group 45° to the left. Move out ³/₄″ toward the tip and bend back 45° to the right.

Bend strip 3 of each group 45° to the right. Move out ³/₄″ toward the top and bend back 45° toward the left. Go all the way around the can, clamping strip 2 of one group to strip 3 of the adjoining group (Fig. 5–7f). Curl strip 2 down to the left and strip 3 down to the right. Curl down tight over clamp (Fig. 5–7g).

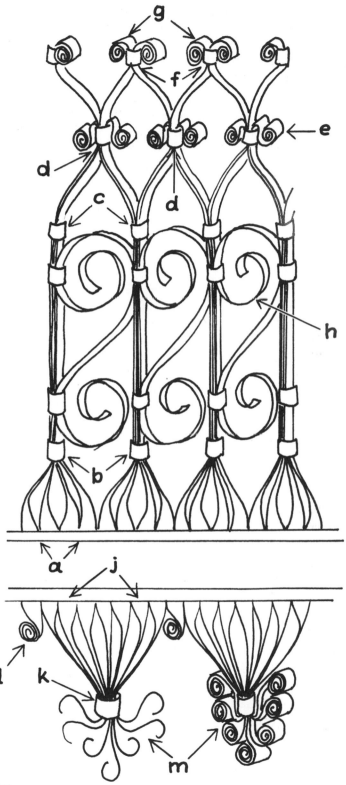

Fig. 5–7   Design for fruit basket

97

The second can will form the legs of the basket. Remove top and top rim. Remove side seam. Shorten can to about $3\frac{1}{2}$ or 4″. (Save this top section—you will need it to make the "question mark" or curlicue inserts shown in Fig. 5–7h.)

Divide can into ten equal sections. Divide each section into eight equal strips. Hold the can bottom-side-up. Start at the space where seam was removed and work left to right. Twist the first four strips one-quarter turn counterclockwise. Twist the next four strips one-quarter turn clockwise. Go all the way around the can, twisting four strips counterclockwise and then four clockwise (Fig. 5–7j).

Clamp the first seven strips together 1″ from bottom of can (Fig. 5–7k). Skip one strip, clamp the next seven together, skip the next one, clamp the next seven, etc., all the way around can. These groups of seven will form the ten legs of the basket. Curl each of the ten single strips down (Fig. 5–7l). Curl each of the strips in the legs down (Fig. 5–7m). Watch Out!—some curl in and some curl out.

Fasten the two cans (basket and legs) together with either solder or glue. Cut eighteen strips of tin $3/16$″ wide (or same width as strips in basket) and 8″ long. Use the excess tin you cut from the second can. If more strips are needed, use a can with the same gauge tin. Form these strips into curlicues (Fig. 5–7h). Clamp them into basket. Paint basket an antique white. Paint curlicues and top curls gold.

HEART BASKET

Materials needed for this project are three $2\frac{1}{8}$″-diameter juice cans and one aluminum beverage can.

The first juice can will become the basket base or legs. Remove top and top rim. Remove side seam. Divide can into eight equal sections (*see* Ch. 1). Divide each section into eight equal strips. Hold the basket bottom-side-up and number the first sixteen strips 1 through 16.

Start at the space where seam was removed and work left to right. Twist strips 1, 2, and 3 one-quarter turn counterclockwise (Fig. 5–8a). Remove strip 4 (Fig. 5–8b). Twist strips 5, 6, 7, and 8 one-quarter turn clockwise (Fig. 5–8c). Clamp these seven strips together $7/8$″ from bottom of can (Fig. 5–8d). Curl them (Fig. 5–8e). This forms the first leg.

Remove strips 10, 12, 14, and 16 (Fig. 5–8f). Shorten strips 9, 11, 13, and 15 about 2″. Twist them one-quarter turn counterclockwise. Curl them down tight to the right (Fig. 5–8g).

Number the next sixteen strips 1 through 16 and repeat the steps in the preceding two paragraphs. Follow this same procedure two more times and the basket legs will be complete.

The second juice can is used to make the top row of added-on hearts. Remove top and top rim. Remove bottom rim. Remove side seam. Flatten out this sheet of tin. Cut twenty-four narrow strips the full length of tin (about $6\frac{1}{2}$″ long). Cut the same width as strips on basket.

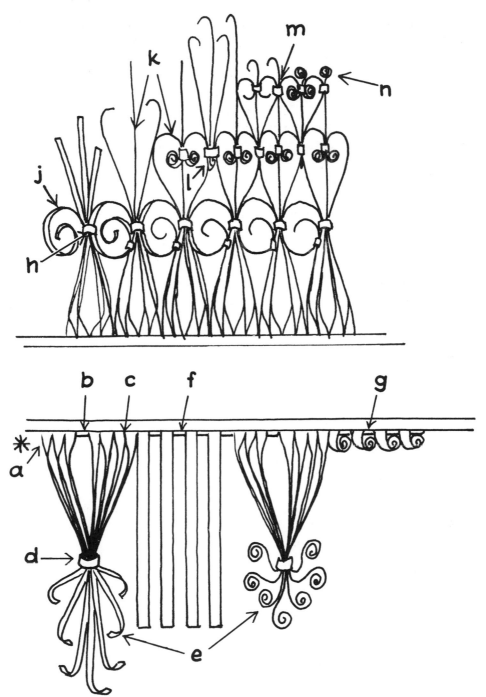

Fig. 5–8   Working pattern for heart basket

Cut eight of these strips in half. *Fold* the remaining sixteen in half. Slip each one of the cut strips down into one of the folded strips. Using your needle-nose pliers, bend about $1/16''$ up on the bottom of each of these sixteen groups (Fig. 5–9). Lay these aside for the time being.

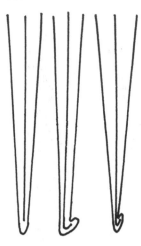

Fig. 5–9   Inserts for heart basket

Remove the top and top rim of the third can. Remove side seam. Divide can into eight equal sections. Divide each section into eight equal strips. Hold can right-side-up and number the strips 1 through 64. Start at the space where seam was removed and work from left to right.

Twist strips 1 and 2 one-quarter turn counterclockwise. Twist strips 3 and 4 one-quarter turn clockwise. Clamp these four together 1″ from bottom of can (Fig. 5–8h). Make this a temporary clamp until you add the large curl from the adjoining group of four strips.

Twist all sixty-four strips, first two counterclockwise, next two clockwise, then clamp in groups of four.

Make a large curl out of the lefthand strip of each group (strips 1, 5, 9, 13, 17, and so on). Clamp these curls in with the group of fours on the left (Fig. 5–8j).

You should now have sixteen groups with three strips each. Curl strips 2 and 4 in a large loop toward strip 3. Clamp 2 and 4 to 3, 1³/₄″ from bottom of can (Fig. 5–8k). Go to the next group of three and make a heart (strips 6 and 8 clamped to strip 7). Go all the way around can and complete the sixteen hearts.

Add a row of hearts on top of the ones you just completed. Take each of the sixteen groups of three (the strips you made from the second can and laid aside) and clamp them one at a time in between the already completed hearts. Clamp 1³/₄″ from bottom of can (Fig. 5–8l).

Form a second row of hearts with this new batch of standees. Form them as before—curl the two outside strips toward the center in loose curls. Clamp them to the center strip 2⁷/₈″ from bottom of can. Clamp the hearts together, with the standing strip from the hearts below between them (Fig. 5–8m).

You now should have thirty-two short strips protruding above the last row of hearts. Curl them all down tight to the right (Fig. 5–8n).

Strips cut from the aluminum beverage can are used to make the basket handle. Remove top and top rim. Cut about ¹/₂″ off top of can, enough to eliminate the neck. As aluminum cans have no seams, you can cut strips as long as you like by starting at the top and going round and round, barber pole fashion (Fig.

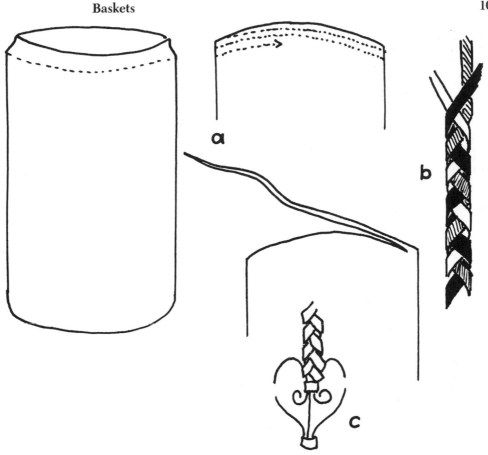

Fig. 5–10   Cutting and braiding a handle for heart basket

5–10a). Keeping your strips about $^1/_{16}''$ wide, take three of these strips, each about 24″ long, and braid them together (Fig. 5–10b). When you have a braid the desired length, attach it to the basket (Fig. 5–10c).

Place the basket between your palms and gently press it in at the sides. This will give your basket a pretty oval shape.

Paint and add flowers.

BIRDCAGE FLOWER BASKET

To complete this basket, shown in color Fig. 8, you'll need one $2^1/_8''$-diameter can; one flat piece of tin, at least $4^1/_2''$ square; one aluminum can to make flowers; fine wire to make flower stems; Styrofoam or florist's clay to hold flowers in basket.

Cut a $4^1/_2''$-diameter disk from a flat piece of tin. Put one point of your compass in center of disk. Put other point of compass 1″ from edge of disk. Draw a circle (Fig. 5–11a). Turn the disk over and draw another circle $^1/_2''$ in from edge of disk.

Divide and mark—*do not cut*—disk into twelve equal pie-shaped sections.

Place your needle-nose pliers along the righthand side of these lines, one at a time, and give a slight clockwise twist. Place your pliers on the lefthand side of each line and give a slight counterclockwise twist.

Be sure the tip of your pliers rests on, but does not cross over, the 1″ circle line as you twist up this design.

Turn disk over and punch a ⅛″ slot on the ½″ circle line, halfway between each of the twelve wedges (Fig. 5–11b). It is easier to assemble basket when these holes have been punched from the underside, rather than punched from the top.

Note: Here are two ways to make a tool to punch a ⅛″ slot: (1) file the tip of a small screwdriver down to a sharp edge; (2) file or grind the tip end of a square-cut nail down to a sharp edge.

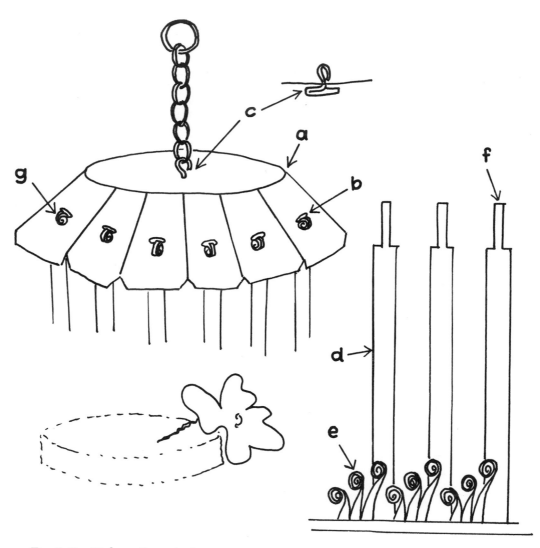

Fig. 5–11   Birdcage flower basket

Punch a small hole in center of disk. Affix a wire so a chain can be attached later (Fig. 5–11c).

Remove top and top rim from the $2^1/_8''$ can. Remove side seam. Draw a line around can $3/_4''$ from top edge. Divide can into six equal sections. Divide each section into four equal strips. This will give you twenty-four strips, approximately $1/_4''$ wide (Fig. 5–11d).

Hold the can in an upright position. Start at the space where seam was removed and number strips 1 through 24. Shorten all odd-numbered strips to $1^1/_2''$ from bottom of can.

Divide each of these twelve short strips into three equal strips. Twist all thirty-six strips one-quarter turn counterclockwise and curl them down (Fig. 5–11e).

Divide and cut the ends of the even-numbered strips into three equal smaller strips. Cut down only to the $3/_4''$ line you drew around the can at the beginning.

Remove the two outside strips on each group of three. Snip them off at the $3/_4''$ line (Fig. 5–11f).

Insert Styrofoam disk or florist's clay in bottom of can.

Attach basket to its top by inserting each strip of the basket through a slot in the top and curling the strip down tight against the top (Fig. 5–11g).

Paint. Add 2″ or 3″ of chain to top. Add flowers.

Chapter 6

Candleholders

SCONCE CANDLEHOLDER

Materials needed to complete this project (Fig. 6–1) are one 1-pint beer can, 2³/₈" diameter by 6¹/₄" tall (not aluminum); a piece of fine wire; paint.

Remove top and top rim; *do not* remove side seam. Draw a line around can 1" from bottom (Fig. 6–1a).

Draw a line 1¹/₄" from seam on each side (Fig. 6–1b). Cut along these lines from the top edge to the 1" mark. This gives you a section 2¹/₂" wide with the seam down the middle.

Shorten the other section to 4¹/₂" from bottom of can (Fig. 6–1c). Divide this section into six smaller equal sections. Cut down to the 1" mark and no further. Now divide each of these six sections into eight equal strips (Fig. 6–1d).

Curl every other strip down. Give each strip exactly two full turns of the curling tool. Curl toward the can. Bring these twenty-four curled ends down to the bottom rim of the can in a graceful arc (Fig. 6–1e).

Cut a piece of fine wire about 12" long. Thread this wire through the twenty-four curled ends. Draw them in tight against bottom rim of can. Twist ends of wire together (Fig. 6–1f).

Cut off 1" of the remaining twenty-four strips and curl them down tight, curling away from the can (Fig. 6–1g).

Cut five ¹/₈" strips down each side of the remaining large section (Fig. 6–1h). Starting at the outside edges, cut the first strips down to within ¹/₂" of the 1" mark. Each succeeding strip should be ¹/₂" shorter (Fig. 6–1j). Shorten each of these ten strips to about 2". Curl them down as shown (Fig. 6–1k).

Shape the remaining standing piece as shown (Fig. 6–1l). Paint.

104

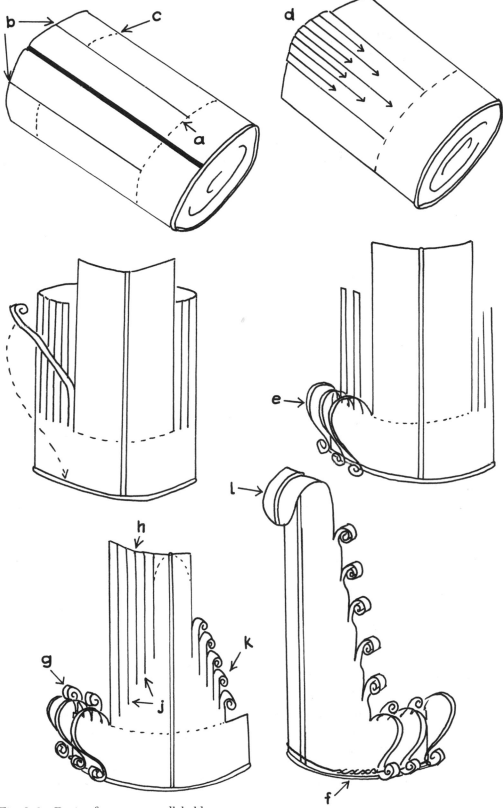

Fig. 6-1   Design for sconce candleholder

105

PIE-PLATE SCONCE

Materials needed to complete this project (Fig. 6–2) are one 1-gallon
motor oil can; one bottom or lid from a 4¼″ can; one bottom or lid
from a 2⅝″ can; three 6 x 32 short bolts.

It is optional whether you use a can bottom or a cut disk from the tin sides. I
prefer the bottoms, as they are a heavier gauge metal and they also have ridges
which give an added design to the finished project.

Remove top and top rim from a 1-gallon motor oil can. Remove bottom and
bottom rim; remove side seam. Trim about ¹/₁₆″ off this bottom disk. (This will
give a safer dull edge to work with.) Place one point of your compass in center of
disk and other point ¾″ from edge and draw a circle. *Do not cut*—just mark disk
into sixteen equal pie-shaped sections (Fig. 6–3a).

Place needle-nose pliers on the righthand side of these lines, one at a time,
and give a slight clockwise twist. Place pliers on the lefthand side of each line and
give a slight counterclockwise twist. Be sure the tip of your pliers rests on, but
does not cross over, the ¾″ circle as you make the above designs.

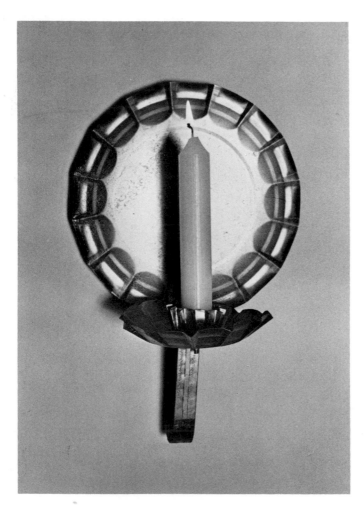

Fig. 6–2   Pie-plate sconce

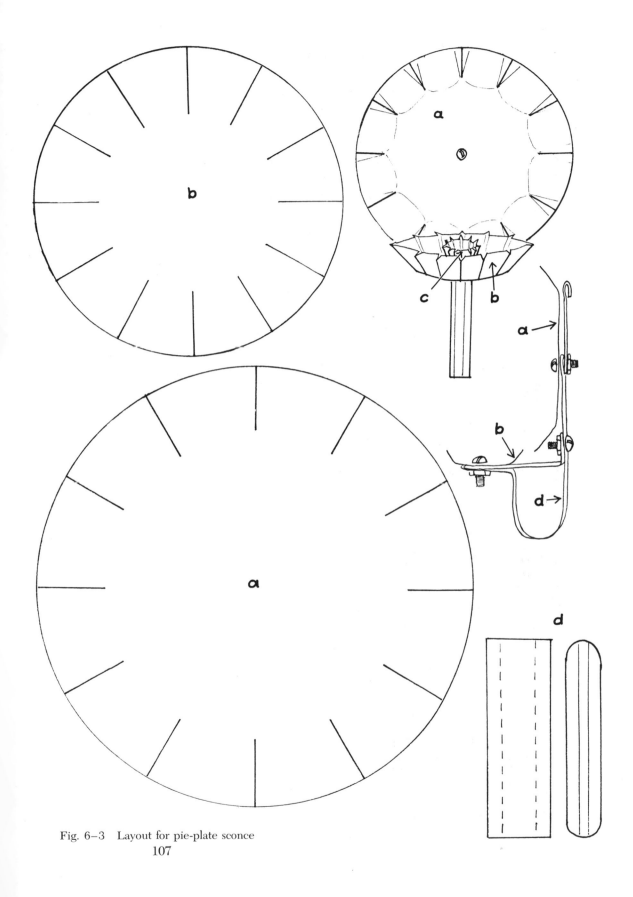

Fig. 6–3   Layout for pie-plate sconce

107

Remove the bottom or lid from a 4¼″ can. Trim about ¹/₁₆″ off the edge of the disk. Place one point of compass in center of disk and the other point ³/₄″ in from outer edge and draw a circle (Fig. 6–3b). Mark—*do not cut*—disk into twelve equal pie-shaped sections. With your needle-nose pliers, shape this disk in the same manner as you did the gallon-can disk.

Place one point of your compass in center of the 2⁵/₈″ disk and other point 1″ in from edge and draw a circle. Mark into twelve equal pie-shaped sections.

With your needle-nose pliers, shape this disk in the same manner as you did the other two, only work the sides of this piece in tight. Make it cup shaped, not saucer shaped as the others (Fig. 6–3c).

From the flat piece of tin (sides of the motor oil can), cut a strip 1″ wide the full 18″ length. Mark a line along the full length, ¹/₈″ in from each edge. Fold these ¹/₈″ sections 180°. That is, bend them right back onto the strip. This will tend to make the strip more rigid (Fig. 6–3d).

Cut a piece of this strip 12″ long. Punch a hole ¹/₂″ from each end. Cut another piece 4¼″ long and punch a hole ¹/₂″ from each end. Bend these strips and bolt them to the disks as shown. Paint.

HURRICANE LAMP

To make this lamp (color Fig. 4), you'll need one 2⁵/₈″ can; one 3¹/₈″ can; one 4¼″ can; one 4″ can; one 6 x 32 bolt; one glass chimney to fit a 3″ base.

Remove top and top rim from 2⁵/₈″-diameter can. Remove bottom and save it. Remove side seam. This piece will form the choker that goes on the throat of the glass chimney. Shorten can to about 3″ from bottom. Divide can into fourteen equal sections. Divide each section into five equal strips. Start at space where seam was removed and go all the way around can, alternately twisting *two* strips clockwise and *three* strips counterclockwise. Gather in groups of five and clamp about ⁵/₈″ above bottom of can. Using a small curling tool, curl all five strips down tight (Fig. 6–4).

The 2⁵/₈″ can bottom that you saved will form the cup that holds the candle. Mark this disk into twelve equal pie-shaped sections. Draw a ³/₄″ circle in center. Use needle-nose pliers and crimp up tight (Fig. 6–5a). Punch a hole in center.

Fig. 6–4 Choker for hurricane lamp

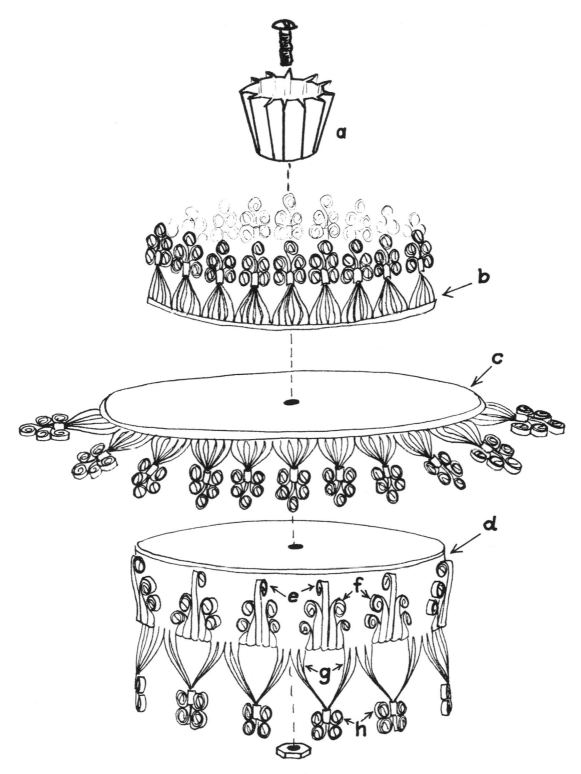

Fig. 6–5   Working pattern for hurricane lamp

Remove top and top rim from 3$^1/_8$″ can. Remove side seam. Shorten can to about 3″ from bottom. Divide can into sixteen equal sections. Divide each section into five equal strips. Twist and curl as you did the first can. This will form the part that holds the glass chimney (Fig. 6–5b). Punch a small hole in center for a 6 x 32 bolt.

Remove top and top rim from the 4$^1/_4$″ can. Remove side seam. Shorten can to about 3″ from bottom. Divide can into twenty-two equal sections. Divide each section into five equal strips. Bend all strips down flat (they will extend from can bottom like spokes from a wheel hub). Twist and curl in groups of five, just as you did for the other cans. This will form the section between the chimney holder and the base (Fig. 6–5c).

Remove top and top rim from the 4″ can. Shorten can to about 4″ above bottom. Draw a line around can about $^3/_4$″ from bottom. Remove side seam only to this line. Divide can into ten equal sections, cutting down to the $^3/_4$″ line. Divide each section into nine equal strips. An easy way to accomplish this is to cut a $^1/_8$″ strip off each section and then cut the remaining section in half. Divide each of these halves into four equal strips. Starting at the space where the seam was removed and working toward your left, number the strips in each group from 1 to 9 (Fig. 6–5d). Curl strip 5 of each group toward the can down to bottom rim (Fig. 6–5e). Cut about $^1/_2$″ off strips 4 and 6 of each group. Curl 4 down toward the can and give a counterclockwise twist as you finish the curl (Fig. 6–5f). On strip 6, give a clockwise twist as you curl down. Cut about $^3/_4$″ off strips 3 and 7. Curl and twist the same as you did strips 4 and 6. They will be a little shorter. Twist strips 1 and 2 of each group one-quarter turn counterclockwise. Twist strips 8 and 9 one-quarter turn clockwise (Fig. 6–5g).

Form strips 1, 2, 8, and 9 of each group around a $^3/_4$″-diameter dowel stick. Clamp these four together and then curl them down (Fig. 6–5h).

Punch small hole in center for bolt. This can forms the base of your hurricane lamp. Bolt four sections together. Paint. Add a string of white beads around the outside of chimney holder if desired.

AMERICANA LAMP

Materials needed to make this project (Fig. 8, color section) are one 1-gallon motor oil can; one 3$^1/_8$″-diameter can; one 2$^5/_8$″ can lid or bottom; one glass chimney with a 3″ base; two 6 x 32 short bolts.

Remove the top and top rim from a clean 1-gallon motor oil can. Remove bottom and bottom rim. Remove side seam.

Be careful of this tin. Can openers have a tendency to leave a sharp and ragged edge, especially on the bottom disk, which you are going to use. If you trim about $^1/_{16}$″ off all the way around the bottom it will make a safer edge to work with.

Draw a 3$^1/_8$″ circle in center of bottom disk. Mark—*do not cut*—into twelve equal pie-shaped sections. Place your needle-nose pliers along the right side of these lines. One at a time, give a slight twist clockwise. Be sure the tip of your

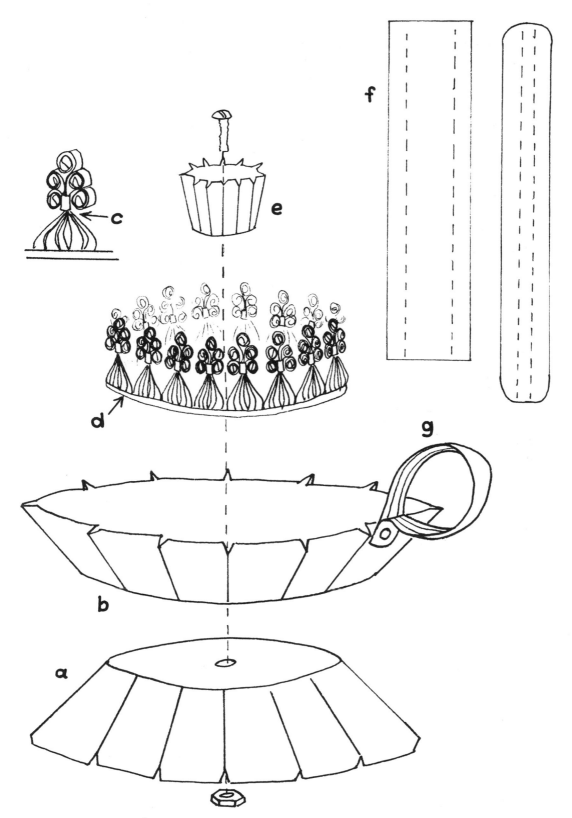

Fig. 6–6   Americana lamp pattern

111

pliers rests on $3^1/_8''$ circle mark. Now put your pliers on the left side of these lines and give a slight twist counterclockwise.

When all twelve have been twisted, punch a small hole in the center just large enough for a small bolt and lay aside (Fig. 6–6b).

Draw and cut a 5″-diameter disk from the flat can side. Place one point of your compass in center of disk and other point 1″ in from edge and draw a circle.

Mark into twelve equal pie-shaped sections. Using your needle-nose pliers, crimp each of these twelve sections in the same manner as the can bottom. This piece will be the stand or base of the candleholder. The crimped sections should be drawn up tight to form a cup, rather than a saucer. Punch a hole just large enough for a small bolt in center (Fig. 6–6a).

The $3^1/_4''$-diameter can should have the top, the top rim, and the side seam removed. Shorten the can to about $2^1/_2''$ from bottom. Divide can into sixteen equal sections. Cut each section into five equal strips.

In each group of five strips, twist two clockwise and three counterclockwise. Clamp these groups of five together about $3/_4''$ from bottom of can (Fig. 6–6c).

Using a small curling tool, curl each of these five strips down to clamp. This forms the section of lamp that will hold the glass chimney in place (Fig. 6–6d). Punch a hole just large enough for a small bolt in the center of it.

The $2^5/_8''$-diameter can lid should be marked off into twelve equal pie-shaped sections. Place one point of your compass in center of disk. Place other point $7/_8''$ from edge and draw a circle. Use needle-nose pliers to crimp up tightly in the same manner as other two disks. This is the cup that will hold the candle (Fig. 6–6e).

Cut a strip of tin from the flat can side, about 1″ wide and 5″ long. This piece will form the handle of the holder (Fig. 6–6f).

Draw a line, $1/_8''$ from each edge, the full length of the strip. Fold these two $1/_8''$ sections 180°, flat over the strip. This will give more strength to the handle. Punch a hole about $1/_2''$ from each end. Install this piece as shown in Fig. 6–6g. Bolt all five pieces together as shown.

CANDLE OR TOOTHPICK HOLDER

For this project (color section, Fig. 4), you need: one $2^1/_8''$-diameter can; one $2^5/_8''$-diameter can; one votive candle glass; one 6 x 32 x $1/_2''$ nut and bolt; fine wire and beads (optional).

Remove top and top rim of the $2^1/_8''$-diameter can. Draw a line around can 1″ from bottom. Remove side seam down to the 1″ mark.

Divide can into eight equal sections (cut down only to the 1″ mark). Divide each section into nine equal strips. To divide the sections into nine strips, I would suggest that you cut one $1/_{16}''$ strip down each section, then divide the remainder into eighths the usual way. It may be less confusing if you cut and shape each section as you go (Fig. 6–7a).

Hold the can right-side-up and number the strips in each section 1 through 9, left to right.

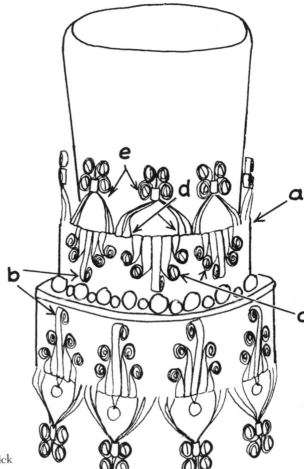

Fig. 6–7   Design for candle or toothpick holder

Snip about ¹/₂″ off strip 5. Place curling tool on end of strip and start curling toward you. Bend strip all the way down, flat against the can. Curl up until curl just reaches the bottom rim (Fig. 6–7b).

Snip about ¹/₂″ off strips 4 and 6. Snip about ³/₄″ off strips 3 and 7. Curl strip 6 down toward you until it is ¹/₄″ shorter than strip 5. Before you remove your curling tool, twist curl one-quarter turn counterclockwise (Fig. 6–7c). Do the same to strip 4, only twist one-quarter turn clockwise.

Curl strip 7 down toward you until it is ¹/₂″ shorter than strip 5. Before you remove your curling tool, twist curl one-quarter turn counterclockwise. Do the same to strip 3, only twist one-quarter turn clockwise. Twist strips 1 and 2 one-quarter turn clockwise. Twist strips 8 and 9 one-quarter turn counterclockwise (Fig. 6–7d).

Place a short length of a ⁵/₈″-diameter dowel stick between strips 1 and 2 and strips 8 and 9. With your needle-nose pliers, pull these four strips together and clamp them (Fig. 6–7e). Curl strips 1 and 2 down to the left and strips 8 and 9 down to the right.

Punch a hole in the bottom center just large enough to take a 6 x 32 bolt. The second can is prepared in exactly the same manner.

Bolt the two cans together. Paint. Add beads around middle to match color of votive glass or candle.

CANDLESTANDS

The directions and patterns for the candlestands (color section, Fig. 4) all call for 4¼″ juice cans. If you would like to use a smaller can, it is a simple matter just to cut the strips a bit narrower.

Stand 1

Remove top rim from a 4¼″ can. Remove side seam. Shorten can to about 3½″ from bottom. Divide can into twelve equal sections. Divide each section into eight equal strips.

Hold the can bottom-side-up and number the first sixteen strips 1 through 16. Start at the space where seam was removed; work from right to left.

Remove strips 1, 3, 5, and 7. Shorten strips 2, 4, 6, and 8 to about 2″ from bottom of can. Twist them one-quarter turn clockwise and curl them down to the right into the empty spaces (Fig. 6–8a).

Twist strips 9, 10, 11, and 12 one-quarter turn clockwise. Twist strips 13, 14, 15, and 16 one-quarter turn counterclockwise. Clamp these eight strips together about 1¼″ below bottom of can (Fig. 6–8b). Curl strips 9 and 16 up over the clamp. (These two strips curl in the opposite direction from the remaining six.) Curl strips 10, 11, 14, and 15 up tight against the two previous curls. Add a clamp to the two remaining strips 12 and 13 (Fig. 6–8c). Place this clamp up tight against curls 11 and 14, then curl strips 12 and 13 up tight.

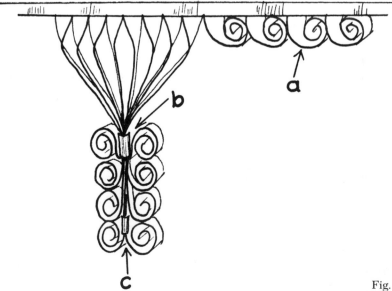

Fig. 6–8   Candlestand 1

You have now completed one section of six. Move to the left and work the next sixteen strips up in the same manner. When this section is complete, move on around and complete all six sections. Paint desired color.

Stand 2

Remove top and top rim from a 4¼″ can. Remove side seam. Shorten can to about 3″ from bottom. Divide can into ten equal sections. Divide each section into eight equal strips.

Hold the can bottom-side-up and number the first sixteen strips 1 through 16. Start at the space where seam was removed and work from left to right.

Shorten the first eight strips to about 2″ from bottom of can. Bend strips 2, 4, 6, and 8 straight up 180°. Curl them away from you, down onto the rim of can (Fig. 6–9a).

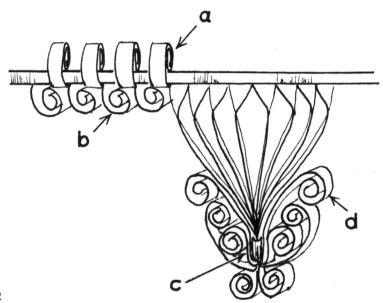

Fig. 6–9 Candlestand 2

Twist strips 1, 3, 5, and 7 one-quarter turn counterclockwise. Curl them down to the right into the empty spaces (Fig. 6–9b).

Twist strips 9, 10, 11, and 12 one-quarter turn clockwise. Twist strips 13, 14, 15, and 16 one-quarter turn counterclockwise. Clamp these eight together about 1¼″ below bottom of can (Fig. 6–9c), then curl them up to form one of the legs. Look at Fig. 6–9d very closely, as some of these strips curl in opposite directions.

This completes one section of five. Move to the right and work the next sixteen strips up in the same manner. Keep moving around until all five sections have been completed.

Stand 3

Remove top and top rim from a 4¼″ can. Remove side seam. Shorten can to about 3″ from bottom. Divide can into twelve equal sections. Divide each section

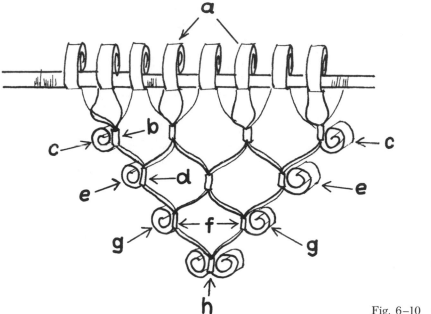

Fig. 6–10   Candlestand 3

into eight equal strips. Hold the can bottom-side up and number the strips 1 through 96, starting at space where seam was removed. Work toward the right.

Bend all the odd-numbered strips straight up 180°. Shorten them to about 2″ from bottom of can. Curl them all away from you, down onto the rim of the can (Fig. 6–10a).

Twist strips 2, 6, 10, and 14 one-quarter turn clockwise. Twist strips 4, 8, 12, and 16 one-quarter turn counterclockwise. Clamp strips 2 and 4 together about ³/₈″ below bottom of can (Fig. 6–10b). Clamp 6 and 8 together; clamp strips 10 and 12 together; clamp strips 14 and 16 together: all are clamped ³/₈″ from bottom of can. Cut about 1″ off the ends of strips 2 and 16 and curl them down over the clamps (Fig. 6–10c).

Clamp strips 4 and 6 together about ³/₈″ below previous clamp (Fig. 6–10d). Clamp strips 8 and 10 together; clamp strips 12 and 14 together: all are clamped about ³/₈″ from previous clamp. Cut about ¹/₂″ off strips 4 and 14 and curl them down over the clamps (Fig. 6–10e).

Clamp strips 6 and 8 together and 10 and 12 together, ³/₈″ from previous clamp (Fig. 6–10f). Curl strips 6 and 12 down over the clamp (Fig. 6–10g).

Clamp strips 8 and 10 together ³/₈″ from previous clamp and curl them down to form the foot of the first leg (Fig. 6–10h). Finish the second leg by forming the next eight strips in a like manner. The last four legs also should be worked up the same way.

Stand 4

Remove top and top rim from a 4¹/₄″ can. Remove side seam. Shorten can to about 3″ from bottom. Divide into twelve equal sections. Divide each section into six equal strips.

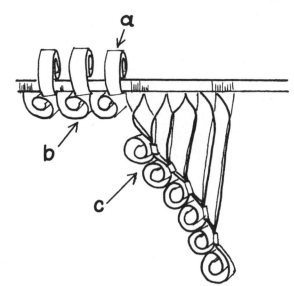

Fig. 6–11 Candlestand 4

Hold can bottom-side-up and number the first twelve strips, 1 through 12. Cut about 1″ off the first six strips. Bend strips 2, 4, and 6 straight up 180°. Curl them down over the rim of the can, curling them away from you (Fig. 6–11a). Twist strips 1, 3, and 5 one-quarter turn clockwise. Curl them down toward the right into the empty spaces (Fig. 6–11b).

Twist strips 7, 8, and 9 one-quarter turn clockwise. Twist strips 10, 11, and 12 one-quarter turn counterclockwise. Clamp strips 7 and 8 together about ¹/₂″ from bottom of can. Curl strip 7 up over the clamp (Fig. 6–11c). Clamp strips 8 and 9 together, just as close to the previous clamp as you can get. Curl strip 9 up over the clamp.

Clamp strips 9 and 10 together, just as close to the previous clamp as you can get. Curl strip 10 up over the clamp. Add strips 11 and 12 to this group, one at a time, in the same manner.

This completes the first section and leg. Continue around the can in a like manner until all six legs are completed.

Stand 5

Remove top and top rim from a 4¹/₄″ can. Remove side seam and shorten can to about 3″ from bottom. Divide can into ten equal sections. Divide each section into eight equal strips.

Hold can bottom-side-up and number the first eight strips 1 through 8. Twist strips 1, 2, 3, and 4 one-quarter turn clockwise. Twist strips 5 through 8 one-quarter turn counterclockwise.

Shorten strip 1 to about 2″ from bottom of can. Curl it down to within about ¹/₂″ from bottom. Curl it to the left (Fig. 6–12a).

Clamp strips 2 through 8 together about 1″ from bottom of can. Curl these seven strips as shown in Fig. 6–12b. Watch closely, as some strips curl in opposite directions.

Fig. 6–12   Candlestand 5

This completes the first leg. Work around can, forming nine more legs in the same manner.

Stand 6

Remove top and top rim from a $4^1/_4''$ can. Remove side seam. Shorten can to about 3″ from bottom. Divide can into ten equal sections; divide each section into eight equal strips. Hold the can bottom-side-up and number the first eight strips 1 through 8.

Twist strips 1, 2, 3, and 4 one-quarter turn clockwise. Twist strips 5, 6, 7, and 8 one-quarter turn counterclockwise. Clamp strips 2 through 8 together $^3/_4''$ below base of can (Fig. 6–13a).

Shorten strip 1 to about 2″ from bottom of can. Curl it down to within $^1/_2''$ from bottom of can (Fig. 6–13b). Curl the seven up as shown in Fig. 6–13c.

This completes the first leg. Work around the can, forming nine more legs in the same manner.

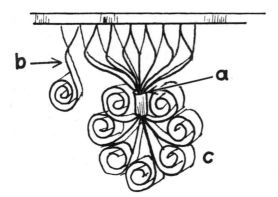

Fig. 6–13   Candlestand 6

Chapter 7
Christmas Ornaments

TWELVE-POINTED MEDALLION

Materials needed to make this project (Fig. 7–1 and color Fig. 10) are one large can lid or bottom, 4″ or more in diameter; one $1\frac{1}{2}″$-diameter Styrofoam ball; beads and sequins.

Fig. 7–1   Christmas ornaments: *from left*, twelve, eight, and sixteen-pointed medallions; sunray medallion and Christmas angel

119

Make sure the edges of your can lid have had all the rough spots removed or flattened out.

Find the center of lid and place one point of your compass in the center. Place the other point $^3/_4$″ out from center and draw a circle; draw a second circle $^1/_2$″ out from the first; draw a third circle $^1/_4$″ out from the second.

Divide can into six equal pie-shaped sections. Cut from outer edge in to the first or inner circle ($^3/_4$″ from center). Make sure your cutters are pointing straight toward the center as you make these cuts. Cut three narrow strips—between $^1/_{16}$″ and $^1/_8$″ wide—to the inner circle, between each of the six sections (Fig. 7–2a). We will come back to these eighteen strips.

Divide the remainder of the six pie-shaped sections in half and cut into the second or middle circle ($1^1/_4$″ from center). Cut three narrow strips between each of these new cuts (Fig. 7–2b). Curl these eighteen strips in toward the center. Use a small curling tool (Fig. 7–2c).

Put a mark at the outer edge in the center of each of the twelve sections (Fig. 7–2d). Place your needle-nose pliers in a line from one of these points to the righthand edge at the intersection of the third or outer circle ($1^1/_2$″ from center). Bend down sharply. Place your pliers in a line from this same point to the lefthand edge at the intersection of the same circle. Bend down sharply. Bend

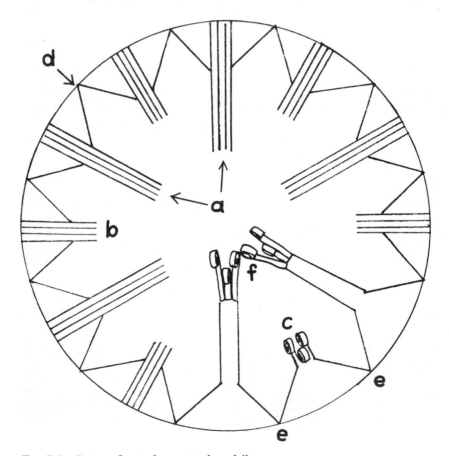

Fig. 7–2   Pattern for twelve-pointed medallion

both of these triangles all the way back onto themselves. This will form a point at the ends of each of the twelve sections (Fig. 7–2e).

Go back now to the six groups of thin strips left near the center. Curl the center strip of each group halfway down, curling toward you. Place your curling tool on the very tip of each of the remaining twelve strips. Curl them toward you, giving your curling tool just one complete revolution (Fig. 7–2f).

Cut the 1½″ Styrofoam ball in half. Place one half of ball in the center of medallion and bend these last twelve strips around it to hold it in place.

Paint medallion desired color and add a few small beads or sequins to Styrofoam ball.

EIGHT-POINTED MEDALLION

To complete this medallion (Fig. 7–1; color Fig. 10), you'll need one lid from a 3⅛″ can; one 1″-diameter Styrofoam ball; a straight pin and beads or sequins.

Find the center of the 3⅛″ can lid and place one point of your compass in the center. Place the other point of your compass ½″ from center and draw a circle. Draw a second circle with point of compass 1″ out from center.

Divide disk into four equal pie-shaped sections. Cut in to the inner circle (Fig. 7–3a). Divide each of the four sections in half and cut in to the outer circle.

Round off the eight sections like petals of a flower (Fig. 7–3b). Place your needle-nose pliers along the line from the end of the short cut to the inner circle and give a slight twist up (Fig. 7–3c).

Cut the 1″ Styrofoam ball in half. Fasten one half of ball in center of medallion by pushing a ¾″ straight pin through the ball and through a small hole in medallion. Bend pin over on back side. Paint if desired and add a few small beads and sequins.

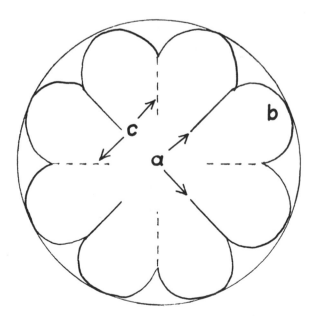

Fig. 7–3   Pattern for eight-pointed medallion

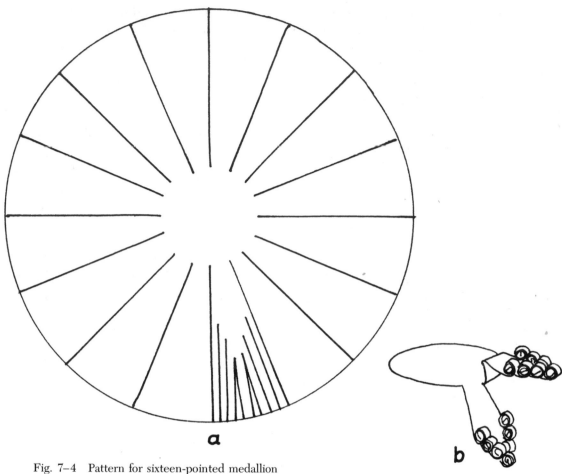

Fig. 7–4   Pattern for sixteen-pointed medallion

SIXTEEN-POINTED MEDALLION

Materials needed for this medallion (Fig. 7–1; color Fig. 10) are one lid or bottom from a 4″-diameter can; one 1″-diameter Styrofoam ball; beads and sequins.

Find center of 4″ lid. Place one point of your compass in the center. Place the other point 1/2″ out from center and draw a circle. Draw a second circle 5/8″ out from the first.

Divide lid into sixteen equal pie-shaped sections. Cut in to the first or inner circle (Fig. 7–4a). Each one of these sixteen sections will end up with seven curls, so try to keep the strips you cut just under 1/8″ wide.

Cut one strip on both sides of each section. Cut in to the second or outer circle (1 1/8″ from center). Use a small curling tool and curl these thirty-two strips in tight toward the center. Curling these strips now will get them out of the way so they don't interfere with the next cutting.

Cut thirty-two more strips, one on each side of each of the sixteen sections. Terminate these cuts about 1/4″ short of first set of cuts. Curl these thirty-two down until they abut the first curls.

Cut the third set of thirty-two strips, just as before. Terminate these cuts about 1/4″ short of last set of cuts. Curl them down until they abut the last set of curls. If the tips of the remaining sixteen strips seem too wide, trim them down a little and then curl them in to abut other curls.

Bend every other section straight up at the inner circle. Bend these same strips back, 1/4″ up from base. This will give your ornament a three-dimensional effect (Fig. 7–4b).

Cut the 1″ Styrofoam ball in half. Glue one half of the ball in center of medallion. Paint if desired and add a few beads and sequins.

SUNRAY MEDALLION

You'll need one lid from a 4″-diameter can and one 1″-diameter Styrofoam ball to make this project (Fig. 7–1; color Fig. 10).

Find center of 4″ lid. Place one point of your compass in center, with the other point of your compass 1/2″ out from center; draw a circle. Draw a second circle with point of compass 1 1/4″ out from center.

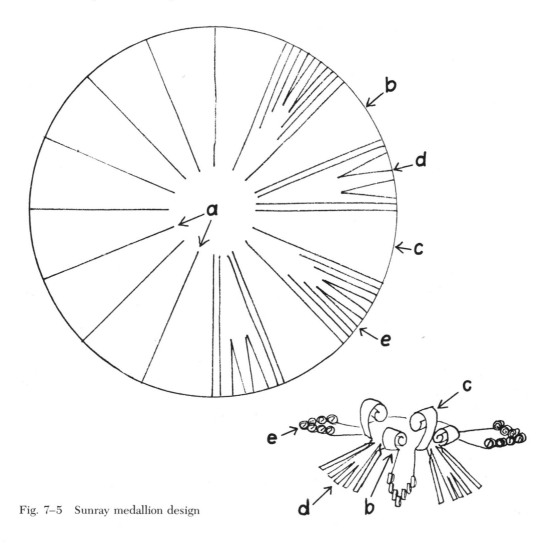

Fig. 7–5 Sunray medallion design

Divide lid into sixteen equal pie-shaped sections. Cut in to the inner circle (Fig. 7–5a). Number the sections 1 through 16. Roll sections 1, 5, 9, and 13 on a pencil toward the center of disk (Fig. 7–5b). With your needle-nose pliers, roll sections 3, 7, 11, and 15 into a larger roll up onto the smaller rolls (Fig. 7–5c).

Divide sections 2, 6, 10, and 14 into five equal strips. Flair them out just a little, fan fashion (Fig. 7–5d).

Cut one strip on both sides of the remaining four sections. Cut in to the second or outer circle. Cut these strips between $1/16''$ and $1/8''$ wide. Use a small curling tool to curl these eight strips in toward the center. Cut eight more strips, one on each side of the four sections. Terminate these cuts about $1/4''$ short of the first set of cuts. Curl these eight down to abut the first set of curls (Fig. 7–5e).

Cut the third set of eight strips just as before. Terminate these cuts about $1/4''$ short of last set of cuts. Curl them down to abut last set of curls.

If the tips of the remaining four strips seem a little wide, trim them down a little and curl them in. Fasten a $1''$ Styrofoam ball in the center and paint.

CAP AND GOWN

Materials needed for this ornament (Fig. 7–6) are one lid from a 4″-diameter can; one lid from a 3″-diameter can; one 2″-diameter Christmas tree ball; one 6 x 32 x $1/2''$ bolt.

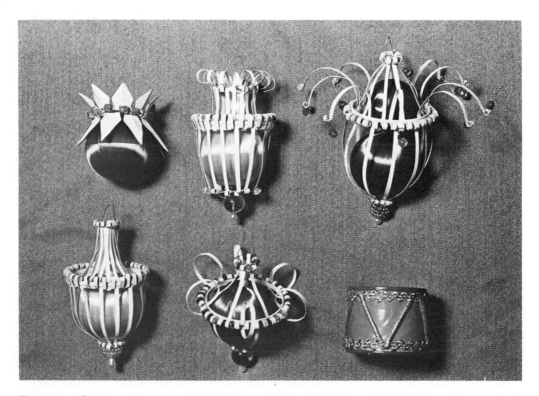

Fig. 7–6   Christmas ornaments: *from left*, cap and gown, birdcage, chandelier, stocking cap, curly top, and drum ornaments

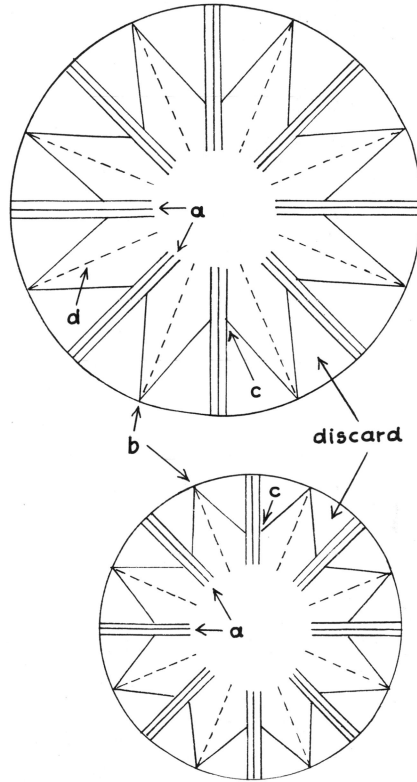

Fig. 7–7   Layout for cap and gown ornament
125

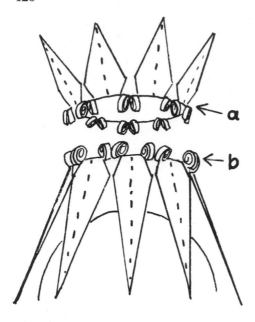

Fig. 7–8   Assembly of cap and gown ornament

Find the center of the 4″ disk. Place one point of your compass in center, the other ⅝″ out from center, and draw a circle. Draw a second circle with point of compass 1″ out from center.

Divide disk into eight equal pie-shaped sections. Cut in to the first or inner circle. Cut sixteen ⅛″-wide strips. Cut one strip down each side of the eight sections (Fig. 7–7a). Use a small curling tool to curl these strips down toward the center (Fig. 7–8b).

Mark the center of each of the eight sections on the outer edge (Fig. 7–7b). Cut from this mark to the point where the outer or second circle intersects with the right edge of section. Cut from this center mark down to where the second circle intersects with the left edge of section (Fig. 7–7c). This will give you an eight-pointed star.

Mark the center of each of the eight sections at the first or inner circle. Place your needle-nose pliers along the line from this center mark out to the outer point of star and give a slight clockwise twist (Fig. 7–7d). Bend all eight points up 45° from the inner circle (Fig. 7–8a).

Follow exactly the same procedures with the 3″ disk.

Bolt the two stars together, bottom to bottom. Paint. Fasten 2″ Christmas ball in place.

BIRDCAGE

Materials needed to complete this ornament are one small juice can, 2⅛″ in diameter, and extra tin, same gauge; one small Christmas ball, 2″ diameter; fine wire.

Remove top and top rim from juice can. Remove bottom and side seam. Divide can into ten equal sections. Divide each section into six equal strips.

Hold the can bottom-side-up. Starting at space where seam was removed, number strips 1 through 60, left to right. Bend every even-numberd strip straight up over the bottom rim. Shorten these strips to 2″. Use a small curling tool and curl each of these strips down to the rim. Curl toward you (Fig. 7–9a).

Work all the way around can, beginning with strip 3, bending up every fourth strip thereafter (3, 7, 11, 15). Put a mark on each of these strips about 1¼″ from rim. Put a small curling tool on each mark and, twisting tool away from you, give one full turn (Fig. 7–9b).

Cut several strips of tin the same width as strips on can. Using the same size curling tool, curl up fifteen small curls (or beads) to be used as spacers between the fifteen upright strips (Fig. 7–9c).

Run a wire through the curls on these fifteen uprights and insert a bead or spacer between each strip. Twist wire ends together when circle is complete. Form the upper ends of these fifteen strips into large graceful curls (Fig. 7–9d).

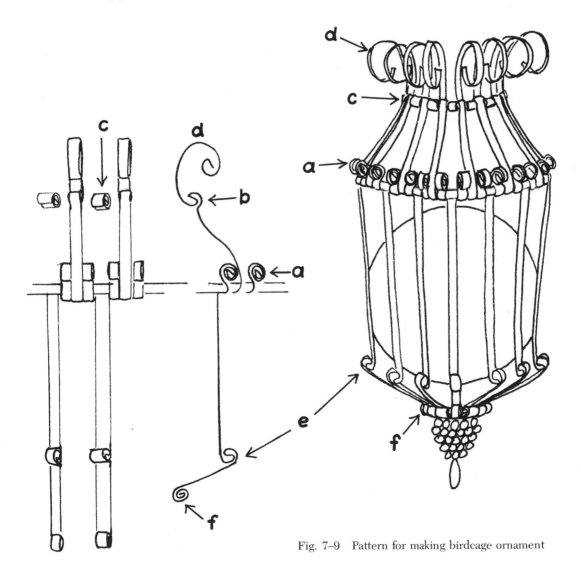

Fig. 7–9   Pattern for making birdcage ornament

Put a mark 2″ from rim on the fifteen remaining strips. Put a small curling tool on each mark and, twisting tool away from you, give one full turn (Fig. 7–9e).

Place curling tool on end of each of these strips and give one full turn away from you (Fig. 7–9f).

Paint. Hang a 2″ Christmas ball in the center. Run a wire through the curls on the bottom and pull them tightly together. Twist ends of wire together to form a circle.

CHANDELIER

To make this project (Fig. 7–6), you'll need one small juice can 2⅛″ in diameter; twelve small 8-mm Christmas balls or twelve large beads; one 2″ satin ball; one 1½″ satin ball.

Remove top and top rim from juice can. Remove bottom and side seam. Divide can into ten equal sections. Divide each section into six equal strips.

Hold the can bottom-side-up. Starting at space where seam was removed, number the strips 1 through 60, left to right.

Bend the even-numbered strips out and shorten them to about 2″ from bottom of can. Curl them away from you, tightly against rim of can (Fig. 7–10a).

Curl the very ends of strips 1, 7, 13, 19, 25, 31, 37, 43, 49, and 55. Twist curling tool away from you, one full turn (Fig. 7–10b).

Thread a piece of fine wire through each of these curls; pull up into a tight circle after 2″ satin ball has been inserted. Twist ends of wire together.

Push the remaining twenty strips up through the center of can. *Do not* make sharp bends in these strips. Let them flow out from the rim in large curves.

Bend strips 5, 11, 17, 23, 29, 35, 41, 47, 53, and 59 in large graceful curls out over the rim of the can. Small ornaments will dangle from these curls (Fig. 7–10c).

Curl the very ends of the remaining ten strips. Twist curling tool toward you, one full turn (Fig. 7–10d). Thread a piece of fine wire through each of these curls. Pull up into a tight circle after inserting 1½″ satin ball. Twist ends of wire together. Paint.

Hang small Christmas balls from overhanging curls, or use beads if desired.

THE STOCKING CAP

You'll need one small juice can, 2⅛″ in diameter, one 2″ satin ball, and fine wire to complete this ornament (Fig. 7–6).

Remove top and top rim from juice can. Remove bottom and side seam. Divide can into six equal sections. Divide each section into eight equal strips.

Hold the can bottom-side-up and number the strips 1 through 48. Start at the space where seam was removed and work from left to right.

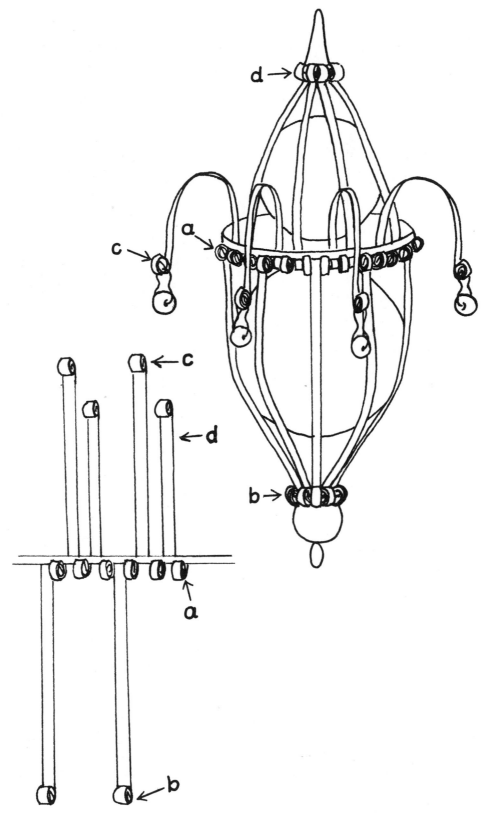

Fig. 7–10   Pattern for chandelier ornament
129

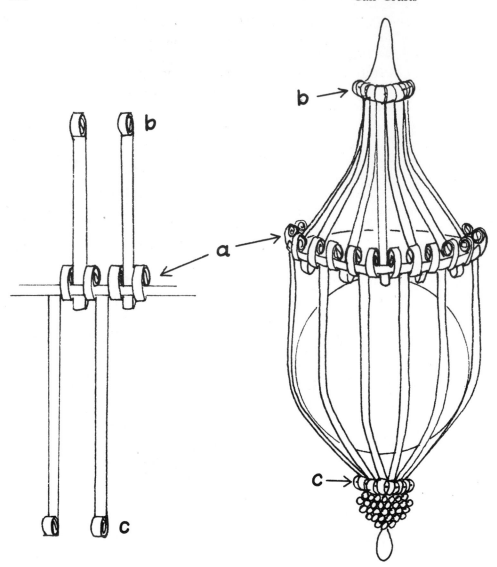

Fig. 7–11   Pattern for stocking cap ornament

Bend all the odd-numbered strips up over the outside of the can rim. Shorten them to about 2″. Curl these twenty-four strips down over the top edge of rim. Curl away from you. As the curls meet the rim, give each one-quarter turn clockwise (Fig. 7–11a).

Bend every fourth strip (2, 6, 10, 14, etc.) up through the center of can, back of the rim. Place a small curling tool on the tip of each of these twelve strips and curl toward you. Make two complete revolutions of curling tool (Fig. 7–11b).

String a fine wire through these curls and pull them into a circle; twist ends of wire together.

Turn can over, bottom-side-down. Place a small curling tool on the tip of each of the remaining twelve strips and curl toward you. Make one complete revolution of curling tool (Fig. 7–11c).

String a fine wire through these curls and pull them into a circle and bind ends of wire together. Before you bind the bottom, you should insert the 2″ satin ball or hang some sort of Christmas ornament inside.

DRUM

Materials needed to make the drum (color section, Fig. 10) are one 6-ounce frozen orange juice can—cardboard, *not* metal; a 7″ piece of red felt ribbon; several feet of gold rickrack.

Remove top from container and save. Use a sharp razor to cut the container off 1½″ from bottom (Fig. 7–12). Glue top back on.

Glue red ribbon around can, then add gold rickrack.

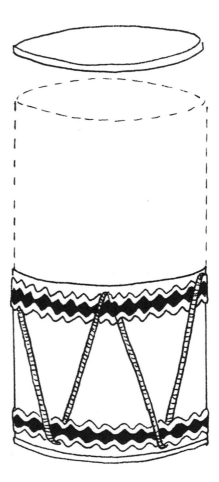

Fig. 7–12   Design for drum ornament

ANGEL

Materials to make the angel, shown in Fig. 7–1, are: one 3″-diameter can lid; one angel head or wooden bead; 3″ of gold rickrack trim.

Fig. 7–13 provides the pattern. After angel has been cut out and painted, add a wooden bead for a head if a craft store head is not available. Glue a piece of rickrack or gold trim along the bottom hem of gown.

Fig. 7–13   Christmas angel pattern

Chapter

Wall Plaques

SUNBURST

To make this plaque (color section, Fig. 11), you'll need two $3^1/8''$-diameter cans, 4″ tall, with ridges; one $5^1/2''$ to 6″-diameter can lid; one 4″-diameter can lid; two 3″-diameter can lids; one $1^1/2''$ flat disk; one 6 x 32 x $1/2''$ bolt.

The materials listed will make a sunburst about 10″ in diameter. If a larger one is desired, use two $4^1/4''$-diameter juice cans, 7″ tall, for the background and work into the center using appropriate size disks.

The two $3^1/8''$ cans with ridges should be identical, since the ridges will show up in the finished project as part of the design. Remove top and top rim of the first can. Mark a line around the can 2″ from the top edge. Remove side seam. Divide into sixteen equal sections.

Stand can on a solid table and bend all sixteen sections or spokes down 90°, flat onto the table. Your can will now resemble a large hub with sixteen spokes reaching out. Divide the end of *each* spoke into five equal strips: Cut the first strip down to the 2″ mark you put on the can before you removed the seam. Cut each of the other three strips $1/4''$ shorter than the previous cut (the last cut should be $1^1/4''$ long).

Using a short length of coathanger wire for a core, coil all five strips down, on each of the sixteen spokes, barber pole fashion (*see* Fig. 1–20). Punch a hole in center of can. Paint black.

Take the second can and remove top and top rim. Mark a line around the can $1^1/2''$ from the top edge. Remove the side seam.

Divide can into sixteen equal sections. Stand can on a solid table and bend all sixteen spokes down 90°, flat onto the table.

Divide the tip of *each* spoke into five equal strips: Cut the first and last strips down to the $1^1/2''$ mark; cut the two center slits $1/2''$ shorter.

Use a medium-size curling tool to curl all five strips on each of the sixteen spokes toward the center of the can, making one complete revolution with the curling tool. Punch hole in center. Paint silver.

133

Now find the center of the $5\frac{1}{2}''$ or $6''$ lid. Place one point of your compass in the center. Place other point of your compass $1''$ in from edge of disk and draw a circle. Punch hole in center.

Cut $\frac{1}{8}''$-wide strips all the way around disk, cutting from the edge in to the $1''$ line. Make sure as you cut that your snips are pointing directly at the hole in center of disk.

Remove every other strip, using your needle-nose pliers.

Grasp the remaining strips at the very ends and twist each three-quarters of a turn in a clockwise direction with your pliers. Paint copper.

Find the center of the $4''$-diameter lid. Place one point of your compass in center. Place other point of your compass $\frac{3}{4}''$ in from edge of disk and draw a circle. Punch a hole in center.

Cut $\frac{1}{16}''$-wide strips all the way around disk, cutting from the edge in to the $\frac{3}{4}''$ line. Do not remove any strips. Grasp each at the very end and twist one-eighth of a turn counterclockwise with your needle-nose pliers. Paint black.

Find the center of the first $3''$-diameter lid. Place one point of your compass in center. Place other point of your compass $\frac{5}{8}''$ in from edge of disk and draw a circle. Punch hole in center. Cut $\frac{1}{16}''$-wide strips all the way around disk, just as before. Make sure your snips are pointing to the center of disk. Do not remove any strips. Grasp each strip on the very end and twist it one-eighth of a turn clockwise with your pliers. Paint silver.

Find the center of the second $3''$-diameter lid. Place one point of your compass in center. Place other point of your compass $\frac{1}{2}''$ in from edge and draw a circle. Draw a second circle $\frac{3}{4}''$ in from edge. Mark the disk into eight equal pie-shaped sections. Cut along these marks to the $\frac{3}{4}''$ circle. Cut three $\frac{1}{16}''$-wide strips down *one* side of each of the eight sections.

Use a small curling tool to curl each of these twenty-four strips in toward the center.

Locate the middle of each of the eight sections (on the edge). Lay your needle-nose pliers along a line from these center marks to the righthand edge where the $\frac{1}{2}''$ circle mark crosses. Bend this part all the way back against disk. Now place your needle-nose pliers along a line from these same center marks to the lefthand edge where the $\frac{1}{2}''$ circle crosses. Bend these parts all the way back against disk. You should now have an eight-pointed star. Paint gold.

Cut a $1\frac{1}{2}''$-diameter disk. Punch hole in center. Paint black.

Assemble, using a 6 x 32 x $\frac{1}{2}''$ bolt.

FIVE-POINTED STAR

Materials needed to make this project (Fig. 8–1) are one $4\frac{1}{4}''$-diameter juice can; one $2\frac{5}{8}''$-diameter can; and one 6 x 32 x $\frac{1}{2}''$ bolt.

Remove top and top rim from large juice can. Remove side seam and shorten can to about $2\frac{1}{2}''$ from bottom. Divide can into sixteen equal sections. Divide each section into eight equal strips. This gives you a total of 128 strips. This design only calls for 125 strips, so three will be eliminated as we work along. Stand the can on a table or some other flat surface. Bend all the strips down 90°, flat onto the table.

Fig. 8–1   Five and eight-pointed stars

Hold the can bottom-side-up. Starting at the space where seam was removed, work toward the right. Number the first twenty-five strips, 1 through 25. Twist the first twelve strips one-quarter turn counterclockwise. Twist the next thirteen one-quarter turn clockwise. Clamp the first three strips together $^3/_8''$ from bottom of can. Clamp the next three groups of three strips each in the same manner. Skip the next strip (13). Then clamp the next four groups together in the same manner.

Fig. 8–2   Details for the five-pointed star

Clamp strips 12 and 14 to strip 13 (Fig. 8–2a). Place this clamp about $7/8''$ from bottom of can. Now clamp strips 11 and 15 to strip 13, $1^5/8''$ from bottom of can. Curl these strips down over the clamp. These three will form the apex of one point of the star. Work the other strips up to this point, as shown in diagram.

Remove one strip and work the next twenty-five the same way. Remember, you only have to remove three strips, so none will be taken out between star points four and five.

Now remove top and top rim from the $2^5/8''$-diameter can. Remove side seam. Shorten can to $1^3/4''$ from bottom.

Divide can into fifteen equal sections. Divide each section into five equal strips. Stand can on table and bend all strips down at a flat 90° angle. Twist all strips one-quarter turn clockwise. Clamp together in groups of five, $3/8''$ from bottom of can. Curl down to make a design as shown in Fig. 8–2. Bolt to center of the first can.

Add a picture or make an aluminum flower for the center of the plaque.

EIGHT-POINTED STAR

To complete the ornament shown in Fig. 8–1, you'll need one $4^1/4''$-diameter juice can; one $2^5/8''$-diameter can; one small bolt.

Remove top and top rim from the large juice can. Remove side seam. Shorten can to about $2^1/2''$ from bottom. Divide can into fifteen equal sections. Divide each section into eight equal strips.

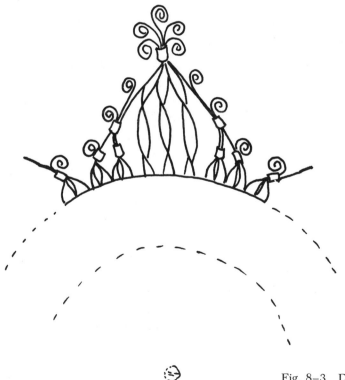

Fig. 8–3   Details for the eight-pointed star

Stand can on a table or some other flat surface and bend all strips down 90°. Hold can bottom-side-up and, starting at the space where seam was removed, number the first fifteen strips 1 through 15.

Remove strips 5, 7, 9, and 11. Twist strips 6, 8, and 10 *one and one-quarter* turns clockwise. Clamp these three together 1½" from bottom of can (this clamp will be temporary).

Twist strips 1, 3, 12, and 14 one-quarter turn counterclockwise. Twist strips 2, 4, 13 and 15 one-quarter turn clockwise. Clamp strips 1 and 2 together, 3 and 4 together, 12 and 13 together, and 14 and 15 together. Place these four clamps ½" from bottom of can.

Remove the temporary clamp and add strips 4 and 12 to the group and reclamp. Curl these five down, as shown in Fig. 8–3.

Curl strip 1 down to the left over the clamp. Clamp strip 2 to strips 3 and 4, ¾" from bottom of can. Curl strip 2 down to the leftover clamp. Curl strip 3 down partway to the left. Work strips 12, 13, 14, and 15 up in a like manner.

Move to the right and work the next fifteen strips up into another point. Continue around the can until all eight points are complete.

Follow directions for five-pointed star to work up the second can (2⅝" diameter), then bolt the two together.

FISH

Materials needed to complete this project (Fig. 8–4) are one aluminum beverage can and one ⅜" eye (available at hobby shops).

Remove top of can and divide can into eight equal sections. Divide each section into eight equal strips.

Stand can up on table or other flat surface and bend all sixty-four strips down 90°, flat onto table.

Start anywhere and remove six consecutive strips (Fig. 8–5a). Hold can bottom-side-up and number the remaining strips 1 through 58, working toward the left. Twist the next five strips (1 through 5) one-quarter turn counterclockwise. Clamp them together about 3" out from bottom of can (Fig. 8–5b)—this will form half of the tail. Curl the ends back to the clamp (Fig. 8–5b). Twist the next five strips (6 through 10) one-quarter turn clockwise. Clamp and curl them the same as the first five strips (Fig. 8–5c).

Remove the next six strips (11 through 16), as shown in Fig. 8–5d. Twist the next ten strips (17 through 26) one-quarter turn counterclockwise. Clamp them together about 2" out from bottom of can. These strips will form the top fin. Curl the ends back to the clamp (Fig. 8–5e). Twist strips 27, 33, 34, and 37 one-quarter turn clockwise (Fig. 8–5f). These will form the upper jaw. Come back to them later. Remove strips 28 through 32, and 35 and 36.

Twist strips 38, 41, 42 and 48 one-quarter turn counterclockwise. These will form the lower jaw.

Remove strips 39, 40, and 43 through 47. Twist the remaining ten strips (49 through 58) one-quarter turn clockwise and clamp them about 2" out from bottom

Fig. 8–4   Butterfly, owl, fish, and rooster wall plaques

of can (Fig. 8–5g). These strips will form the bottom fin. Curl the ends back to the clamp.

Go back and shape the jaws, as shown in Fig. 8–5h. Glue a pop-top ring on the back for a hanger.

OWL

Materials needed to make the owl in Fig. 8–4 are one aluminum beverage can, two eyes, and a nose.

Remove top of can. Divide can into eight equal sections. Divide each section into eight equal strips.

Stand can on table or other flat surface and bend all sixty-four strips down 90°, flat onto table.

Start anywhere and remove ten consecutive strips (Fig. 8–6a). Number the remaining strips 1 through 54. Holding the can bottom-side-up, working from left

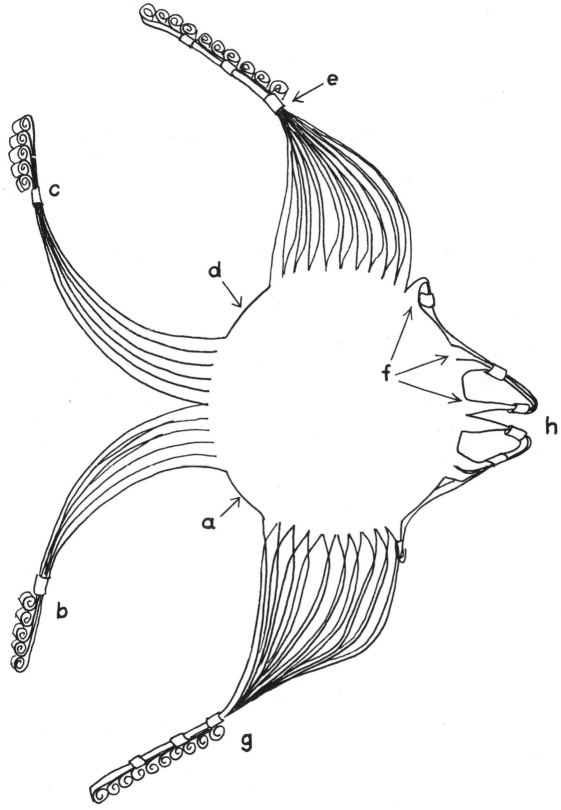

Fig. 8–5　Pattern for fish

139

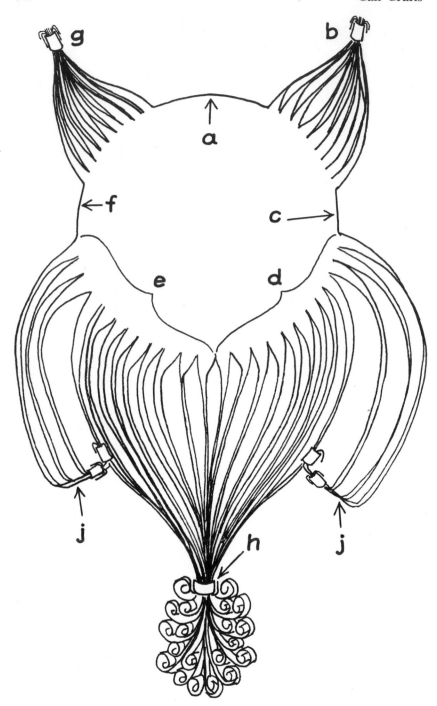

Fig. 8–6   Owl design

to right, twist strips 1 through 9 one-quarter turn clockwise. Clamp these nine together about 1¼″ from rim of can. This will form the first ear (Fig. 8–6b).

Remove strips 10, 11, 12, and 13 (Fig. 8–6c). Twist the next fourteen strips (14 through 27) one-quarter turn clockwise (Fig. 8–6d). (Note: The drawing only shows twelve strips, but fourteen is correct.) Twist the next fourteen strips (28 through 41) one-quarter turn counterclockwise (Fig. 8–6e).

Remove strips 42 through 45 (Fig. 8–6f). Twist the last nine strips (46 through 54) one-quarter turn counterclockwise and clamp them about $1^1/4''$ from rim of can (Fig. 8–6g). This will be the second ear. Finish it off as you did the first one.

Clamp strips 17 through 38 together, just as far out on the tips as you can go. Bend $1/8''$ of the two outside strips back over the clamp to hold it in place (Fig. 8–6h). Curl the remaining strips up to form the owl's tail.

Bring strips 14, 15, and 16 down and form the right wing. Bring strips 39, 40, and 41 down to form the left wing (Fig. 8–6j). Paint and add eyes and nose— or just paint them in. Glue a pop-top ring on the back for a hanger.

ROOSTER

To make the rooster shown in Fig. 8–4, you'll need one aluminum beverage can.

Remove top of can. Divide can into six equal sections; divide each section into eight equal strips.

Stand can on table or other flat surface and bend all forty-eight strips down flat at a 90° angle.

Start anywhere and remove seven consecutive strips (Fig. 8–7a). Hold can bottom-side-up and number the remaining strips 1 through 41, working toward your left.

Twist strips 1, 2, and 3 one-quarter turn clockwise. Twist strips 4 and 5 one-quarter turn counterclockwise. Clamp these five together $1^1/2''$ from bottom of can (Fig. 8–7b).

Twist the next six strips (6 through 11) one-quarter turn counterclockwise. Clamp them together about $1^1/4''$ from bottom of can (Fig. 8–7c).

These first eleven strips will form the rooster's neck and head. Shape them as shown in Fig. 8–7d.

Remove strips 12, 13, 14, and 15 (Fig. 8–7e). Twist strips 16, 17, and 18 one-quarter turn clockwise. Twist strips 19, 20, and 21 one-quarter turn counterclockwise. Clamp these six together $2''$ below rim of can (Fig. 8–7f). These will form one leg.

Twist strips 22, 23, and 24 one-quarter turn clockwise. Twist strips 25, 26, and 27 one-quarter turn counterclockwise. Clamp these six together $2''$ below rim of can (Fig. 8–7g). These will form the second leg. Shape the feet as shown.

Remove strips 28, 29, 30, and 31 (Fig. 8–7h).

Twist the next ten strips (32 through 41) one-quarter turn counterclockwise. These strips will form the tail. Shape them up as shown.

Glue a pop-top ring on the back for a hanger; paint.

Fig. 8–7   Rooster layout

BUTTERFLY

You'll need one aluminum beverage can to make the butterfly (Fig. 8–4).

Remove the top and divide can into nine equal sections. Divide each section into five equal strips. Stand can on a table or other flat surface and bend all forty-five strips down 90°, flat onto table.

Start anywhere and remove one strip (Fig. 8–8a). Hold can bottom-side-up; number the remaining strips 1 through 44, working toward the left.

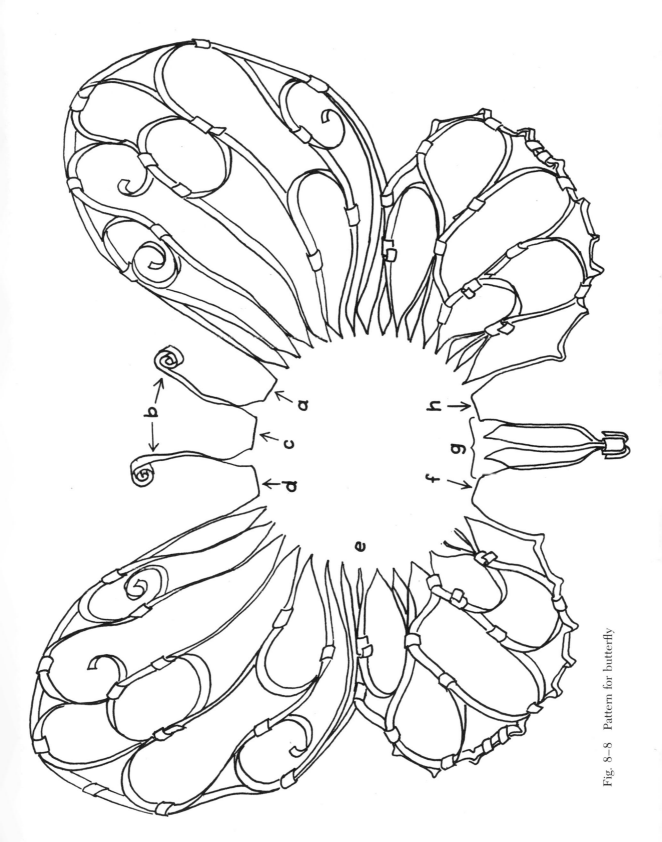

Fig. 8–8 Pattern for butterfly

Twist strip 1 one-quarter turn clockwise. Curl it down to the right, to within 1½″ of bottom of can (Fig. 8–8b). Remove strips 2 and 3 (Fig. 8–8c). Twist strip 4 one-quarter turn counterclockwise. Curl it down to the left, to within 1½″ from bottom (Fig. 8–8b). These two strips (1 and 4) will be the antennae.

Remove strip 5 (Fig. 8–8d). Twist the next sixteen strips (6 through 21) one-quarter turn clockwise (Fig. 8–8e). These sixteen will form a wing: Shape wing as shown.

Remove strips 22 and 23 (Fig. 8–8f). Twist strips 24 and 25 one-quarter turn clockwise. Twist strip 26 one-quarter turn counterclockwise. Clamp these three together about 1½″ from bottom of can (Fig. 8–8g). These three will form rear body of the butterfly.

Remove strips 27 and 28 (Fig. 8–8h). Twist the next sixteen strips (29 through 44) one-quarter turn counterclockwise. These sixteen strips will form the second wing. Shape them just as you did the other wing.

Glue a pop-top ring on back for a hanger. Paint.

Chapter

Cans on Parade

CIRCUS WAGON

Materials needed to complete the wagon, shown in the color section (Fig. 5), are: four soup cans, $2^5/8''$ in diameter for wheels; two 1-gallon rectangular cans, $4^1/8''$ x $6^1/2''$; two 5" lengths of $^1/8''$ aluminum clothesline wire for axles; one flat piece of tin from a soda pop can for the driver's seat.

Wheels: To make *each* wheel, remove top and top rim from one soup can. Remove bottom and side seam. Shorten can to about $2^1/2''$ from bottom. Mark and divide can into eight equal sections. Divide each section into eight equal strips.

All the work to make these wheels will be on the inner side of the can rim. As this space is limited, I would suggest that you push all the strips out, away from the center of can. Do not make any sharp bends, just push them out and away.

Hold the can right-side-up and number the strips 1 through 64, starting at the space where seam was removed and working to your right. One at a time, bend each of the odd-numbered strips toward the center. Using your needle-nose pliers, give each a twist one-quarter turn counterclockwise. Use a small curling tool and curl them down tight to the right (Fig. 9–1a).

Bend strips 2 and 4, 18 and 20, 34 and 36, and 50 and 52 in toward the center. Twist each one-quarter turn clockwise.

Bend strips 6 and 8, 22 and 24, 38 and 40, and 54 and 56 in toward the center. Twist each one-quarter turn counterclockwise.

Clamp strips 2, 4, 6, and 8 together (Fig. 9–1b). Clamp strips 18, 20, 22, and 24 together. Clamp strips 34, 36, 38, and 40 together. Clamp strips 50, 52, and 56 together. Place these clamps about $^5/8''$ from rim of can.

Curl the two outside strips of each group down over the clamp. That would be strips 2 and 8, 18 and 24, 34 and 40, and 50 and 56. Curl the righthand strip of each group (now two strips) down to within $^1/4''$ of first curl (strips 6, 22, 38, and 54). This will be a temporary measure.

145

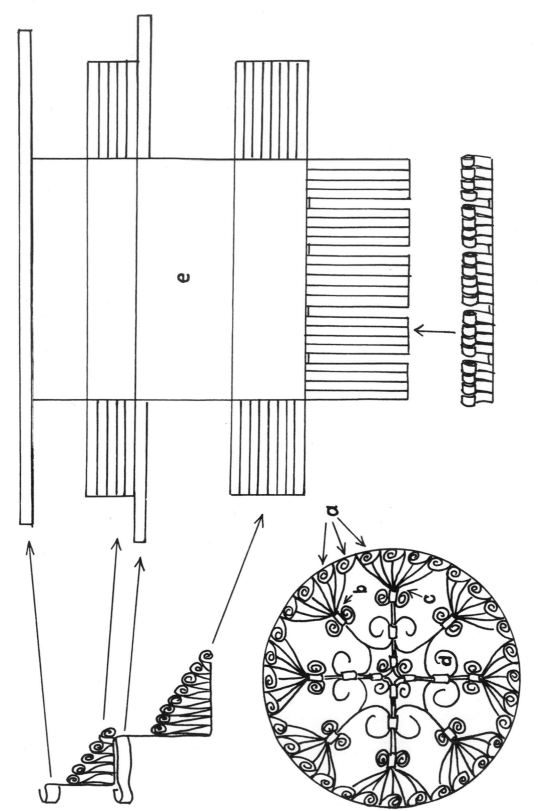

Fig. 9-1 Wheel and seat layout for circus wagon

Draw a circle, the same size as can, with your compass. Put a distinct mark in the center of this circle.

Place the can on the circle and, where the remaining four strips cross the center point, bend them sharply to the left 90°. Snip off about ½" of each of these four and then fit them between the two strips where they now lay. Run the four temporary curls you made a few moments ago back up to the center and clamp.

Bend strips 10 and 12, 26 and 28, 42 and 44, and 58 and 60 toward the center. Then twist each of them one-quarter turn clockwise. Bend strips 14 and 16, 30 and 32, 46 and 48, and 62 and 64 toward the center. Then twist each of them one-quarter turn counterclockwise.

Clamp strips 10, 12, 14, and 16 together (Fig. 9–1c). Clamp strips 26, 28, 30, and 32 together. Clamp strips 42, 44, 46, and 48 together. Clamp strips 58, 60, 62, and 64 together. Place these clamps about ⅝" from bottom of can. Curl the two outside strips of each group down over the clamp. That would be strips 10 and 16, 26 and 32, 42 and 48, and 58 and 64.

You now have two strips left in each of the four groups. Curl one to the left and one to the right (Fig. 9–1d). *Do not* use a curling tool; make large graceful curls. Clamp these curls to the adjoining spokes. Make four of these wheels, then lay aside until final assembly.

Wagon: The two 1-gallon cans will make the wagon. Remove top and top rim from one of these cans. Shorten can to 5" from bottom. Locate the four corners and cut a line from top edge to bottom. Divide the two wide (side) panels into eight equal sections. Divide each of these eight sections into six equal strips. Divide the narrow (end) panels into four equal sections. Divide each of these four sections into six equal strips.

One of these end sections will contain the seam. Remove the seam and divide the remainder of the section into five strips. This end of the can or wagon should now have twenty-three strips (Fig. 9–3). The opposite end should have twenty-four and each side should have forty-eight (Fig. 9–2a).

Hold can right-side-up; the narrow end of can that has the space where the seam was removed should be facing you. Number these strips 1 through 23. If you cut your can the same as I did, you will have two strips to the left of the space where seam was removed (Fig. 9–3a). These will be strips 1 and 2. Twist them one-quarter turn clockwise. Bend strips 3 and 5 down 180°, shorten them to about 2", and curl them up to the left against the bottom rim of can. Twist strip 4 one-quarter turn counterclockwise.

Clamp strips 1, 2, and 4 together ¾" from bottom of can (Fig. 9–3b). Bend strips 7, 9, and 11 down 90° and twist them one-quarter turn counterclockwise. Bend strips 13, 15, and 17 down 90° and twist them one-quarter turn clockwise. Clamp these six together 1" from bottom of can (Fig. 9–3c). This will be the wagon tongue. Shorten strips 7 and 17 to about 2" from clamp and curl them back tightly over the clamp.

Twist strips 6 and 8 one-quarter turn clockwise. Twist strip 10 one-quarter turn counterclockwise. Clamp these three together ¾" from bottom of can. Twist strips 12 and 14 one-quarter turn clockwise. Twist strip 16 one-quarter turn counterclockwise. Clamp these three together ¾" from bottom of can.

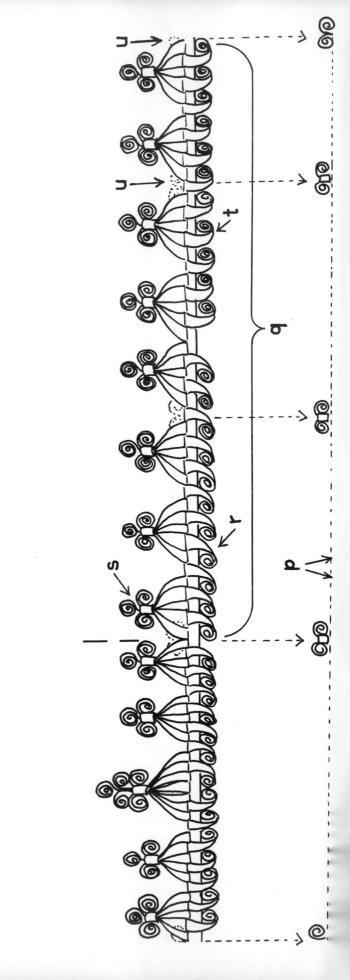

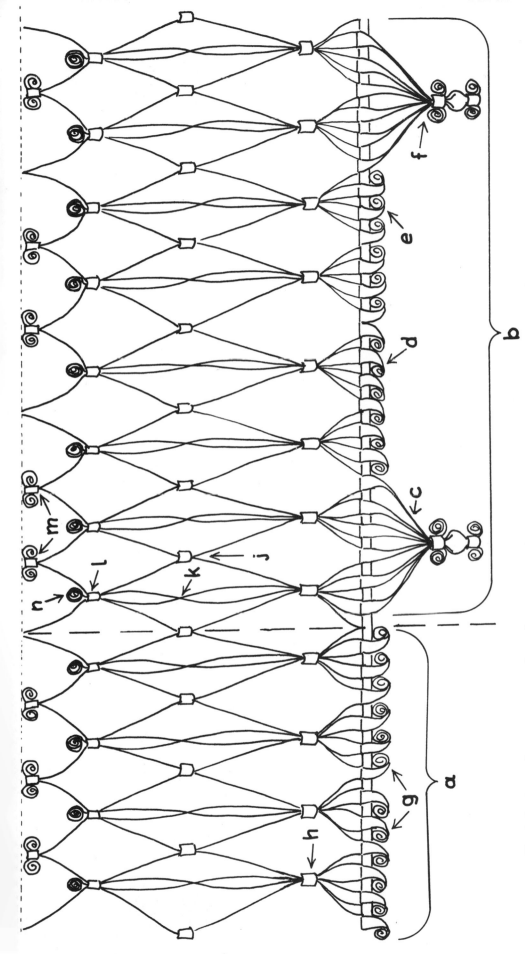

Fig. 9-2 Design for circus wagon sides and top

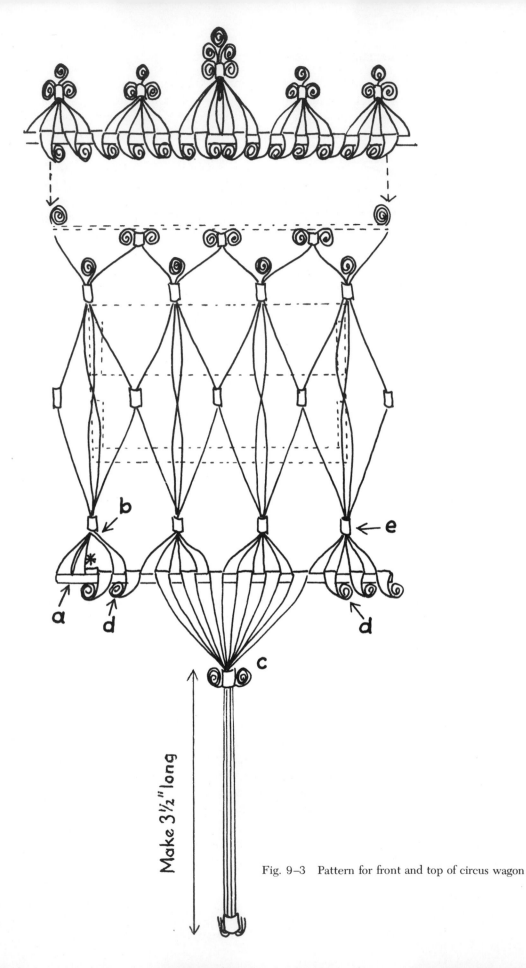

Make 3½" long

Fig. 9–3 Pattern for front and top of circus wagon

Bend strips 19, 21, and 23 down 180°; shorten them to about 2″ from bottom of can. Curl them up, to the right, tight against bottom rim of can (Fig. 9–3d).

Twist strips 18 and 20 one-quarter turn clockwise. Twist strip 22 one-quarter turn counterclockwise. Clamp these three together ³/₄″ from bottom of can (Fig. 9–3e).

Now let's work down one side. Number these strips 1 through 48 (Fig. 9–2b). Bend strips 2, 4, 6, 8, 10, and 12 down 180°. Twist strips 2, 4, and 6 one-quarter turn counterclockwise. Twist strips 8, 10, and 12 one-quarter turn clockwise. Clamp these six together ³/₄″ from bottom of can: This group of six will form one of the wheel holders. (Fig. 9–2c).

Shorten strips 2 and 12 to about 2″ from clamp and then curl them up tightly over the clamp. Shorten strips 4 and 10 to about 2¹/₂″ from clamp and then curl them up over the first two curls. Come back and attach the axles and wheels in the final assembly.

Bend strips 14, 16, 18, 20, 22, and 24 down 180° (Fig. 9–2d). Shorten them to about 2″ from bottom of can. Twist them all one-quarter turn counterclockwise and then curl them up tight, to the left, against the bottom rim of can.

Watch your numbers now! Bend strips 25, 27, 29, 31, 33, and 35 down 180°. Shorten them to about 2″ from bottom of can. Twist them all one-quarter turn clockwise and curl them up tight, to the right, against bottom rim of can (Fig. 9–2e).

Bend strips 37, 39, 41, 43, 45, and 47 down 180°. Twist strips 37, 39, and 41 one-quarter turn counterclockwise. Twist strips 43, 45, and 47 one-quarter turn clockwise. Clamp these six together ³/₄″ from bottom of can. Shorten strips 37 and 47 to about 2″ from clamp and curl them up tightly over the clamp (Fig. 9–2f).

Go to the far side of the can and number the strips 1 through 48. Follow exactly the same procedures as you did for this side.

Now go to the back of the wagon and number the strips 1 thru 24, working from left to right (Fig. 9–2a).

Bend strips 1, 3, 5, 7, 9, and 11 down 180°. Shorten them to about 2″. Twist them all one-quarter turn clockwise and curl them up tight, to the left, against the bottom rim of can.

Bend strips 13, 15, 17, 19, 21, and 23 down 180°. Shorten them to about 2″. Twist them all one-quarter turn counterclockwise and curl them up tight, to the right, against the bottom rim of can (Fig. 9–2g).

Twist strips 2 and 4 one-quarter turn clockwise. Twist strip 6 one-quarter turn counterclockwise. Clamp these three together ³/₄″ from bottom of can (Fig. 9–2h). Repeat for the next 2 groups of 3 each.

Twist strips 20 and 22 one-quarter turn clockwise. Twist strip 24 one-quarter turn counterclockwise. Clamp these three together ³/₄″ from bottom of can. You should now have twenty-four groups of three each around the wagon.

To work the sides up, we will start in the lefthand corner of either side. Clamp strips 5 and 7 together, clamp strips 11 and 13 together, clamp strips 17 and 19 together: All clamps should be 2″ above bottom of can (Fig. 9–2j).

Twist strips 3, 9, and 15 (the center strip in each group) one full twist, barber pole fashion (Fig. 9–2k).

Clamp strips 1 and 5 together with strip 3 between them. Clamp strips 7 and 11 together with strip 9 between them. Clamp strips 13 and 17 together with strip 15 between them. All clamps should be 3″ from bottom of can (Fig. 9–2l).

Clamp strips 5 and 7 together, clamp strips 11 and 13 together, clamp strips 17 and 19 together: All clamps should be 4″ from bottom of can (Fig. 9–2m).

Curl strips 3, 9, and 15 down into the V (Fig. 9–2n). Curl strips 5, 11, and 17 down to the left, over the clamp. Curl strips 7, 13, and 19 down to the right over the clamp (Fig. 9–2p). Follow this same pattern and routine all the way around can. Make and install driver's seat (see Fig. 9–1e).

The second 1-gallon can will form the top to the circus wagon. Remove top and top rim. Shorten can to about 2½″. Locate and cut a slit at each corner. Cut three strips on each side of these first corner cuts. You can probably get only two strips between corner cut and seam. Divide each of the large panels into five equal sections. Divide each of these sections into eight equal strips. Divide the end panels into three sections, then divide each section into eight strips.

Starting at the corner where seam was removed, work to your right, down the long side of the can. Number the strips 1 through 46 (Fig. 9–2q). Start your numbers at the corner cut, not the seam. Bend strips 2, 4, 6, 8, 10, 12, 14, 16, 18, 20, and 22 down 180°. Snip ½″ off the end of each one. Twist them all one-quarter turn clockwise. Curl them all up tight against the bottom rim, curling to the left (Fig. 9–2r).

Twist strips 1 and 3 one-quarter turn clockwise. Twist strip 5 one-quarter turn counterclockwise. Clamp these three together ⅝″ from bottom of can. Then curl the three ends down (Fig. 9–2s).

Twist strips 7 and 9, 13 and 15, and 19 and 21 one-quarter turn clockwise. Twist strips 11, 17, and 23 one-quarter turn counterclockwise. Clamp these strips together in groups of three and curl same as strips 1, 3, and 5. Watch your numbers now!

Bend strips 25, 27, 29, 31, 33, 35, 37, 39, 41, 43, and 45 down 180°. Snip ½″ off the end of each one. Twist them all one-quarter turn counterclockwise (Fig. 9–2t). Curl them up tight against the bottom rim, curling to the right.

Twist strips 24 and 26, 30 and 32, 36 and 38, 42 and 44 one-quarter turn clockwise. Twist strips 28, 34, 40, and 46 one-quarter turn counterclockwise. Clamp these strips together in groups of three and curl, just as before.

This will bring you up to the next corner cut, or the short end of the can. You should have thirty strips between here and the next corner cut. Number them 1 through 30.

Bend strips 2, 4, 6, 8, 10, 12, and 14 down 180°. Snip ½″ off the end of each one. Twist them all one-quarter turn counterclockwise. Curl them all up tight against the bottom rim, curling to the right.

Twist strips 1, 3, 7, and 9 one-quarter turn counterclockwise. Twist strips 5 and 11 one-quarter turn counterclockwise. Clamp these together in groups of three and curl down, as you did on the side.

Twist strips 13 and 15 one-quarter turn counterclockwise. Twist strips 16 and 18 one-quarter turn clockwise. Clamp these four together ¾″ from bottom of can and curl them down.

Bend strips 17, 19, 21, 23, 25, 27, and 29 down 180°. Snip ½″ off the ends of each one. Twist them all one-quarter turn clockwise. Curl them to the left, tightly against the bottom rim.

Twist strips 20, 22, 26, and 28 one-quarter turn counterclockwise. Twist strips 24 and 30 one-quarter turn clockwise. Clamp them together in groups of three and curl down.

This completes one half of the top for the circus wagon. Repeat these same directions for the second half.

If you have a soldering iron and know how to use it, why not solder two strips into the space where seam was removed? It is not necessary but it will make a neater looking finished product.

Assembly: Lay the top onto the wagon proper. Punch small holes in the four corners where the corner posts touch the top. Uncurl the ends of these corners. Insert them through the holes in the top and recurl them down onto the top (Fig. 9–2u). This will hold the top in place on the wagon.

Cut four ³/₈″ disks or washers from a scrap piece of tin. Punch holes in the centers just large enough to slip in the ⅛″ aluminum wire axles. Put wheels on the ends of axles, then slip on the washers. Then peen the ends down with ball-peen hammer.

Lay the axles on the wheel holders between strips 6 and 8. Clamp these strips around axle and curl ends back over the clamps.

WHEELBARROW

Materials needed to make this project (color section, Fig. 9) are: one 1½-pound ham can; two 2⅛″-diameter juice cans for wheels; one 1-quart oil can; five tops and/or bottoms of soup cans to make cups to hold flowers; several aluminum beverage cans for flowers; several yards of wire for flower stems; five 6 x 32 x ½″ bolts.

Barrow: Remove side seam from the 1½-pound can. Divide can into twenty-four equal sections. Divide each section into five equal strips. Hold the can bottom-side-up and work toward the right.

Starting at space where seam was removed, twist two strips one-quarter turn counterclockwise. Twist the next three strips one-quarter turn clockwise. Clamp these five together ½″ from bottom of can. Go all the way around can, twisting and clamping in groups of five. You should have twenty-four groups of five strips each: Groups 1 and 2 on one side of the space where seam was removed; groups 23 and 24 on the other side (Fig. 9–4a). Curl the five strips in each of these four groups down over the clamps. Bend these four groups down 90°, even with the bottom of can. Hold the can bottom-side-down. Number the strips in each group 1 through 5. In groups 3 and 22, curl four strips down tightly (Fig. 9–4b). In the remaining groups, curl strips 1 and 5 down loosely (Fig. 9–4c); this is a move to get them out of the way for the time being.

Place a clamp on the remaining three strips 1⅛″ above bottom of can (Fig. 9–4d). Curl strip 2 down tight to your left over this clamp. Bend strip 4 to your

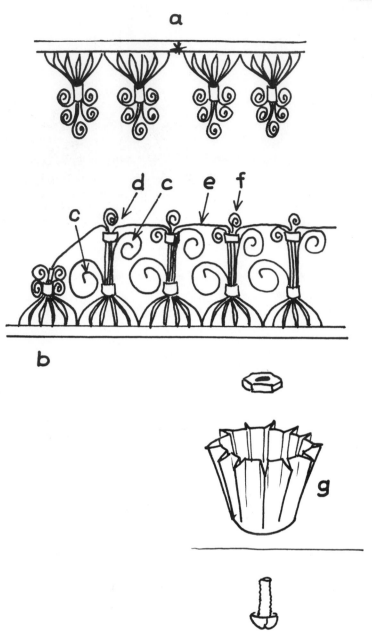

Fig. 9–4   Wheelbarrow: pattern for sides and flowerpot

right 90° (Fig. 9–4e); this strip will fasten into the group on the right. Curl strip 3 down tightly (Fig. 9–4f). Work each group up in the same manner, all the way around can. Now go back and work up a design with the loose curls 1 and 5.

Remove top and top rim from a 1-quart oil can. Remove bottom and bottom rim and remove side seam. Lay this sheet of tin down flat.

Place ham can in center of this sheet and draw an outline of it (Fig. 9–5a). Cut it out as shown and bolt it to the bottom of the ham can in five places. (These bolts will also hold the flowerpots in place when wheelbarrow is finished.)

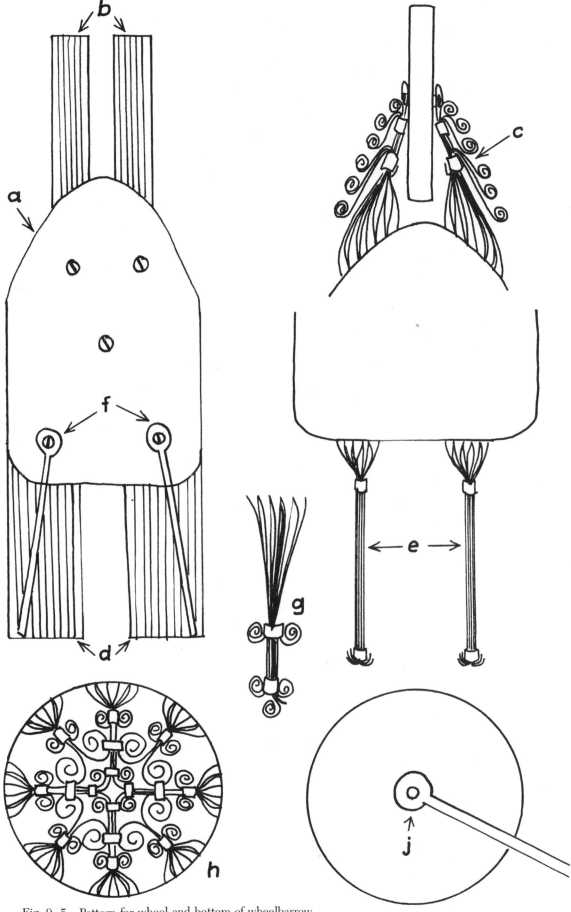

Fig. 9–5   Pattern for wheel and bottom of wheelbarrow

On the wheel end, leave about a $1/2''$ space between two $1''$ sections. Divide each of these sections into five equal strips (Fig. 9–5b).

Twist the five on the right side one-quarter turn counterclockwise and clamp them together about $3/4''$ from rim of can (Fig. 9–5c). This clamp will be temporary until wheel is attached.

Twist the five strips on the left side one-quarter turn clockwise and clamp them together $3/4''$ from rim of can.

On the handle end, leave a $1''$ space between two $13/4''$ sections. Divide each of these sections into ten equal strips (Fig. 9–5d). The five outside strips on each side will bend down 90° to form the legs and the five inside strips on each side will form the handles (Fig. 9–5e).

Twist the five strips on the left leg one-quarter turn clockwise. Add a sixth strip, as shown in Fig. 9–5f, to bolt up onto bottom to add support to the leg.

Clamp these six strips together $11/4''$ below rim of can. Curl the two outside strips up over the clamp. Add another clamp to the remaining four strips $3/4''$ below the first clamp. Curl the strips back over the clamp (Fig. 9–5g).

Twist the five strips on the right leg one-quarter turn counterclockwise. Add a sixth strip and work it up in the same manner as the other leg.

Twist the five strips for the left handle one-quarter turn clockwise. Clamp them together $3/4''$ from rim of can. Add another clamp $2''$ out from the first one. Snip all the strips off $1/8''$ from the last clamp and bend them back over the clamp.

Wheels: Remove the top, top rim, bottom, and side seam from the two $21/8''$-diameter cans. Working with one at a time, shorten can to about $2''$ from bottom. Divide the can into eight equal sections. Divide each section into eight equal strips.

Hold the can right-side-up and number the strips 1 through 64. Start to the right of the space where seam was removed.

Remove strips 1, 2, 3, and 6. Twist strips 4 and 5 one-quarter turn clockwise. Twist strips 7 and 8 one-quarter turn counterclockwise. Clamp these four together $3/8''$ from rim. Snip about $1''$ off strips 4 and 8 and curl them down tightly over the clamp (Fig. 9–5h).

Remove strips 9, 10, 11 and 14. Twist strips 12 and 13 one-quarter turn clockwise. Twist strips 15 and 16 one-quarter turn counterclockwise. Clamp these four together $3/8''$ from rim. Snip about $1''$ off strips 12 and 16 and curl them down tightly over the clamp.

Remove strips 17, 18, 19, and 22. Twist strips 20 and 21 one-quarter turn clockwise. Clamp these four together $3/8''$ from rim. Snip about $1''$ off strips 20 and 24 and curl them down tightly over the clamp.

Work your way around the can in a like manner until all eight sections are complete. Stand the can on a table and, with your thumbs, push all eight sections in toward center of can (90°). Finish the wheel up as shown.

Make two of these wheels and solder them together.

Assembly: Cut about a $3/4''$ length of $1/8''$-thick aluminum clothesline wire for an axle. Push this axle through the center of the wheel and attach a hub on each side. Cut the hubs to shape, as shown in Fig. 9–5j. Peen the ends of the wire down onto the hubs. Clamp the hub ends into the group of five strips on the front of barrow.

Use the tops or bottoms of soup cans to make the five flower holders or cups (Fig. 9–4g). Find the centers of the disks. Place one point of your compass at each center point; place other point of your compass $1/2''$ out. Draw a circle. Divide each disk into twelve equal pie-shaped sections (just mark—do not cut). Place needle-nose pliers along the righthand side of each of these lines, one at a time, and give them a clockwise twist. Make sure the nose of your pliers does not cross over the circle you drew in the middle. Place pliers on the lefthand side of each mark and give a counterclockwise twist. Draw all twelve crimps up tightly to form a cup.

Fasten these cups to the barrow with the bolts that are already in place. Fill these cups with floral clay. Make several small flowers from aluminum cans for each cup. Paint.

FLOWER CART

To complete the flower cart shown in Fig. 9, color section, you'll need: one 1-gallon rectangular can (anti-freeze, Coleman fuel); two $4^1/4''$-diameter cans, for wheels; seven tops or bottoms from soup cans to make cups to hold flowers; nine small bolts; one coathanger; one $6''$ length of $1/8''$ aluminum clothesline for an axle; several aluminum beverage cans for flowers; several yards of fine wire for flower stems; floral clay.

Wheels: Each $4^1/4''$-can will make one wheel (Fig. 9–6a). Remove top and top rim. Remove bottom and side seam. Shorten can to about $4''$ from bottom. Divide can into twelve equal sections. Divide each section into eight equal strips. This will give you a total of ninety-six strips.

Hold the can in an upright position and number the first twenty-four strips, 1 through 24, starting at space where seam was removed and working left to right.

Bend the twelve odd-numbered strips 90° toward the center of the can. Shorten each of these strips to $2''$ and twist them one-quarter turn counterclockwise. Use a small curling tool and curl them in tight to the inside of the rim.

Remove strips 2 and 4. Twist strips 6 and 8 one-quarter turn counterclockwise. Twist strips 10 and 12 one-quarter turn clockwise. Clamp these four together $1''$ from rim of can. Curl the two outside strips (6 and 12) down tight over the clamp. Clamp the remaining two strips (8 and 10) together $2''$ from rim of can. This will be a temporary clamp.

Remove strips 14 and 16. Twist strips 18 and 20 one-quarter turn counterclockwise. Twist strips 22 and 24 one-quarter turn counterclockwise. Clamp these four together $1''$ from rim of can. Curl the two outside strips (18 and 24) down tight over clamp.

Number the next twenty-four strips 1 through 24 and follow the same procedure. When this group is completed, move on to the next twenty-four strips, then the remaining twenty-four strips, and do likewise.

Fig. 9–6 Flower cart: design for wheels and sides

You should now have eight groups with four strips each. Using your right thumb, push groups 1, 3, 5, and 7 in toward the center 90°. These groups will form the hub of the wheel.

Bend strip 10 of each group 90° to the right and insert it between strips 8 and 10 of the next group. You will have to remove the temporary clamp in order to do this. Replace clamp and curl strip 8 down over the clamp. Push groups 2, 4, 6, and 8 in 90° toward the center. Curl and clamp these as shown.

Make two wheels. Use a 5½″ piece of ⅛″ aluminum clothesline wire for an axle. Force the wire through the center of each wheel and peen the ends down with a hammer. Lay aside for the time being.

Main Body: Remove top and top rim from rectangular gallon can. Remove side seam. Do not shorten this can now, as you will need strips of varying lengths. Shorten strips as needed. Locate each of the four corners and cut a slit from the top edge down to the bottom of can.

Divide the end panels into four equal sections. Divide each of these sections into eight equal strips. Divide the side panels into six equal sections. Divide each of these sections into eight equal strips.

Hold can bottom-side-up (if your can is the same as the one I used, the seam you removed was in one corner). Start at this corner and work from left to right. Number the strips 1 through 48.

Bend strips 1, 3, 5, 7, 10, 15, 18, 20, 22, 24, 26, 28, 30, 32, 34, 36, 38, 40, 42, 44, 46, and 48 straight up 180°.

Remove strips 4, 8, 11, 14, 17, 23, 27, 31, 35, 39, 43, and 47. Shorten strips 2, 21, 25, 29, 33, 37, 41, and 45 to about 2″ from bottom of can. Twist them all one-quarter turn counterclockwise. Curl them to the right into large curls, but do not use curling tool (Fig. 9–6b).

The remaining six strips (6, 9, 12, 13, 16, and 19) will form the brace that holds the large wheels (Fig. 9–6c).

Twist strips 6, 9, and 12 one-quarter turn counterclockwise. Twist strips 13, 16, and 19 one-quarter turn clockwise. Clamp these six together about 1⅝″ from bottom of can. Cut strips 12 and 13 off about 1″ below this clamp. Bend them around the aluminum wire axle (do not leave the axle in place at this time, just shape the strips around it), and back under or into the clamped group. Cut strips 6 and 19 off ⅛″ below clamp and bend them back over the clamp (Fig. 9–6c). Curl strips 9 and 16 around a length of coathanger wire (*see* Fig. 1–20). Make these barber pole twists about 2¾″ long. Cut the coathanger wire off and leave it as a core to strengthen these braces. Fasten them to bottom of can, along with the two braces from the other side when it is finished.

Work up the opposite side of can in a like manner, with three exceptions: first, work from right to left, or back to front; second, remove strip 2 (instead of 4) and curl strip 4 (instead of 2); third, the large curls go toward the left.

On the back of the can, number the strips 1 through 32 (Fig. 9–7a). Remove strips 3, 5, 9, 10, 13, 14, 17, 20, 21, 24, 25, and 29. Bend strips 2, 4, 6, 8, 12, 16, 18, 22, 26, 28, 30, and 32 straight up, 180°. Shorten strips 1, 7, 11, 15, 19, 23, 27, and 31 to about 2″ from bottom of can. Twist them all one-quarter turn counterclockwise, and curl them in large curls down toward the right (Fig. 9–7b).

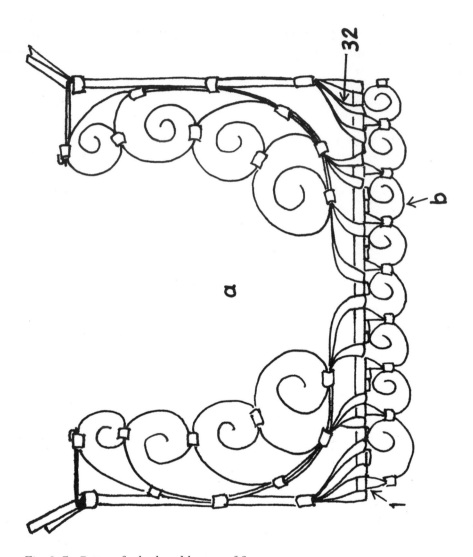

Fig. 9–7   Pattern for back and bottom of flower cart

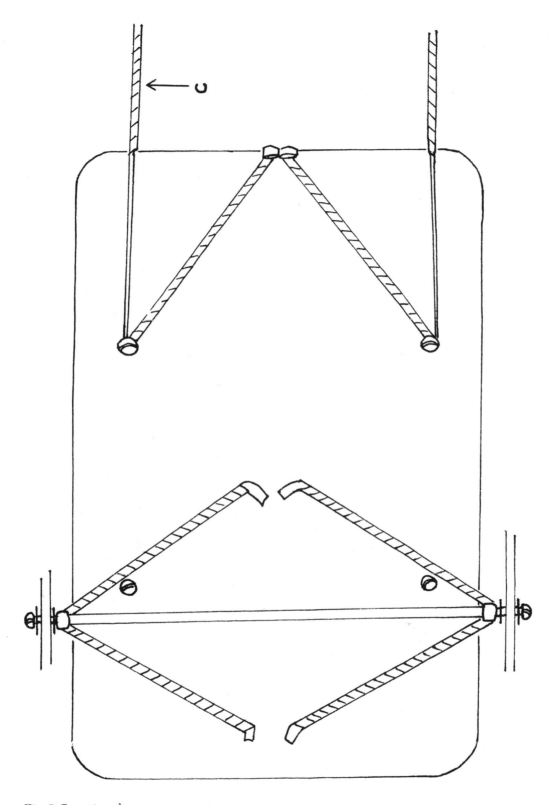

Fig. 9–7 continued.

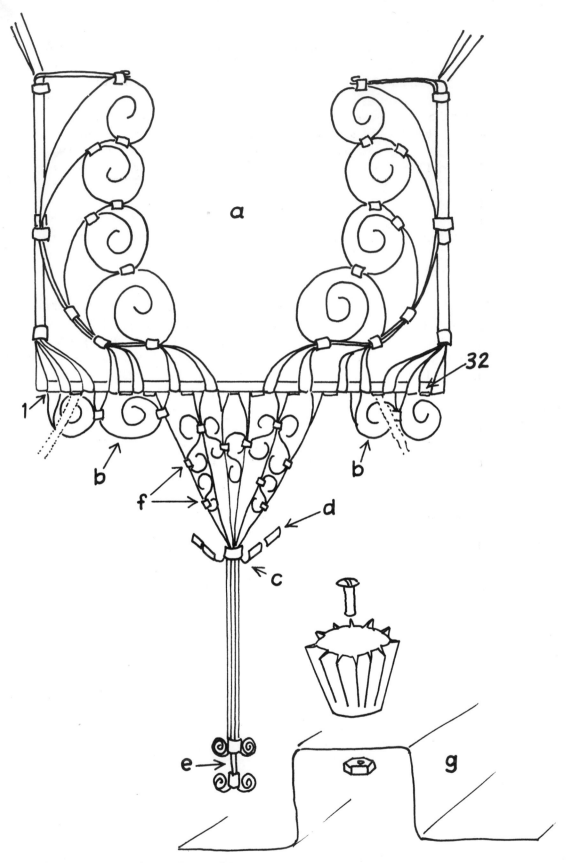

Fig. 9-8   Design for cart front and flowerpot

Fashion sewing

by the

Bishop method

Edna Bryte Bishop

Originator and Developer of the Bishop Method

Marjorie Stotler Arch

Associate Educational Director, Advance Pattern Company

Pattern Sketches by Dorothy L. Davids
Line drawings for Chapters
1 and 2 by Anna R. Atene

Photographs by Stewart Love

J. B. Lippincott Company *Philadelphia • New York*

CONTENTS

Library of Congress Catalog No.: 6Z–16378

PREFACE

This new book, *Fashion Sewing by the Bishop Method*, is designed to help everyone who sews achieve the quality look in all home sewing. The homemaker who sews to find a creative outlet, or who sews to stretch the family budget, now can be sure of doing a professional-looking job by learning the principles of The Bishop Method and by applying them to everyday clothing problems.

The authors' first book, *The Bishop Method of Clothing Construction*, set forth the following fundamental principles of clothing construction: grain perfection; accuracy in preparing, cutting, and marking fabric; cutting to fit; perfection in stitching; perfection in pressing; and correct trimming detail. These basic principles were developed in detail to enable all who used the first Bishop-Arch book to improve their standards of sewing workmanship.

In this second book by the same authors, *Fashion Sewing by the Bishop Method*, the construction principles of the first book are used as the foundation for improving the specific techniques and special skills essential in dealing with the many and varied sewing problems related to clothing all members of the family. A study of

the Table of Contents will reveal the challenging nature and the broad scope of the sewing projects covered.

This new book is not a revision of, nor a replacement for, the first book. There is a real need for both books. You must have the first one to master thoroughly the basic learnings, which, incidentally, are not repeated in this book. You will want the second one to extend these basic learnings and to apply them successfully in the development and improvement of your own sewing techniques.

The important part that sewing plays in modern living calls for a new type of book, one which will meet present-day sewing needs and standards. The girl or woman who sews today has good taste, wants to choose her own pattern and fabric, insists on a style closely identified with her personality, and will settle for nothing less smart than the blouse, dress, or suit which she sees in the best shops. *Fashion Sewing by the Bishop Method* will help every girl and woman attain her individual sewing goals.

In using the book, even though you may not follow through in making all of the garments in any given chapter, you should study the entire chapter to learn all the principles of the Bishop method involved in making garments of each particular type.

In schools, the book can be used from junior high-school level all the way through college level. The chapters used must be chosen to meet the needs, interests, and abilities of the group.

We especially want to give credit and to express our sincere thanks to J. L. Hudson Co. of Detroit, and to Royers in Greensburg, Pennsylvania, for their cooperation with the chapter on alterations of ready-to-wear. Our thanks, also, are extended to Dorothy Davids, artist, and Stewart Love, photographer, for their untiring efforts in helping us to complete the book, and to Joseph Horne Co. of Pittsburgh.

Whether you approach sewing as an art or as a necessity, we present this second book in the hope that you will receive as much enjoyment in using it as we did in writing it for you.

Edna Bryte Bishop
Marjorie Stotler Arch

The well-dressed woman

Being well dressed is one of the finest arts in the world, and there are many things to know and learn in developing this art in ourselves. After you have studied the two chapters on dressing smartly, perhaps you will be able to develop a working plan by listing the *no's* and *must's* that apply to *you*.

Developing the art of being well dressed

You should begin by learning the difference between good and bad taste. Good taste in dress is not a question of money but rather a question of knowledge. What is good taste? It is an expression of personal style to suit your own physical proportions, your personality, and your way of life. It is the ability to judge what is beautiful, appropriate, and harmonious for you. You must learn to distinguish real beauty and elegance from the tricky or eye-catching fashion so frequently seen around us. You may often have to determine your purchases quite apart from personal preference, and test them by the same criteria you would use in viewing a painting or a work of sculpture. Fashion is truly a visual proof of taste.

Then, if taste is what you choose, style is how you wear it. Together, they make elegance in *you*. No amount of wealth can buy style. It takes self-respect, self-discipline, and a good sense of values. Some people have said that style results more from the way one's clothes are worn than from the clothes themselves. It is an elegance which reflects individual personality and vitality. These things are important because your appearance is the way you present yourself to the world.

Here are some points to remember:

1. People with good taste realize the importance of wearing the appropriate clothes for the right occasion. Perhaps, you should begin by analyzing how many sides there are to your life around which to plan your clothes—home, business, sports, social life, travel, and many others.

4. Clothes that are right for you last through the years. Be careful about buying *extreme* fashions, especially in large or expensive items. These may last only one season. A good coat should last five years.

5. Aim to develop a first rate *you*; if fashion seems to be dictating anything that isn't right for you, ignore it. We say that discipline in clothes is a clearcut picture of your own identity as a person. Use fashion instead of fashion using you.

6. Build your wardrobe around never-out-of-season clothes, known as classics.

7. An approach toward complete costume planning is the dress and jacket concept—informal with the jacket on, formal with the jacket off. This will, of course, depend upon the style and fabric from which they are made.

2. Next, before buying anything, review your wardrobe, discarding items you are not likely to wear again. Make a careful wardrobe plan and never, as you shop, depart from your plan by buying a tempting "pretty" that you might or might not ever wear. Choose things that *go together*. Determine one or two basic colors around which your wardrobe will be built, together with two or three colors for accessories, shoes, hats, and gloves. Remember that a man's point of view about clothes is concerned with how pleasing a picture you will make; he will look at the whole ensemble rather than at its separate parts.

You might ask yourself these questions before you make a purchase:

> How will it wear?
> Will it keep its shape?
> Will it become dated?
> Where will it take me?
> How does it fit into my life?

3. Garments such as coats, suits, and dresses, which are worn longest and are most expensive, should be purchased first, then used as a nucleus around which other wardrobe needs are harmonized.

8. The word "functional" has dictated much in American fashion; but too often, it has been interpreted to mean casual, and casual often becomes *too casual*. One example is the number of tourists who visit our nation's capital in shorts and slacks. Complement a public building by your choice of clothes, for you have already learned that your appearance is the way you present yourself to the world. It is not alone just what you wear, but where you wear it.

9. Do not appear on city streets in flat shoes, head scarfs, huge skirts, strapless dresses, or with your hair in pin curls.

10. Ladylike may be the nicest adjective you can merit.

11. The style, color, and fabric in a costume must be exactly right for you. You will learn much about them in these two chapters. Then, just as they play an important role, so do fit and workmanship, because inferior workmanship and poor fit can never be concealed.

12. Do not ever hesitate to wear a becoming costume many, many times.

13. You should have just *one* outstanding or important thing in a costume at a time, such as a beautiful needlepoint bag, a lovely embroidered jacket, or a distinctive hat. Yet, every costume you wear should have something intriguing about it. Judge how you will look twenty feet away!

14. There should never be more than one focal point to a costume. If you wear a brightly colored coat, you must wear a black dress.

15. Many women realize the importance of having something to draw attention to them.

16. Remember that too many spots of contrast are never good in a costume. You must not look cluttered, faddish, or disorganized. For example, on some people, the contrast in a costume is better in the hat and jewelry or collar to form a framework for the face. White gloves moving around a dark costume detract from the face.

17. Check for balance in your costume; for example, the hat which may have a large flower on one side should never be worn when the dress is designed with a large drape on the same side.

18. Always dress yourself carefully, even at home, and remember that women were meant to be *feminine*. Comfortable clothes must never be careless clothes.

19. Now, the foundation has been laid for developing good taste and style to give elegance, charm, true chic, and distinction. These are worth striving for.

Let us analyze the separate parts of a costume.

Choice of accessories

The right accessories can make a costume or ruin it, and they may cost more than your dress, suit, or coat. The most perfect dress, suit, or coat cannot stand on its own without the right wearer and the right accessories.

Accessories will change a costume, so it is wise to select some dresses, suits, and coats you can dress up *or* dress down. Consider a gold pin versus a rhinestone one, a black jersey sash versus a black satin one, and a large leather bag versus a small satin clutch bag.

Accessories must be appropriate. Never wear fancy shoes with a sports costume, long, dangle earrings for traveling, or a business-like leather or straw bag with a filmy frock.

Also, have only one outstanding or important thing in a costume at a time, such as a hat with brightly colored flowers, a plaid suit, or a leopard muff. Yet, every costume you wear should have something intriguing about it.

Don't wear *everything* at once. It is a great American fault. We refer you to the all-important point system for dressing smartly on page 192, *The Bishop Method of Clothing Construction*.

It may take some time to build up a wardrobe of the right accessories, but good leather bags, gloves, and shoes are marks of a well-dressed person.

Hats. 1. A hat is usually the most important accessory to a smart appearance. It lifts an individual into another status and is a symbol of a woman.

2. Did you ever see a suit in a display window without a hat? The suit or the ensemble is incomplete without the right hat.

3. A hat should be ladylike, with great dignity and charm.

4. Develop a flair for wearing hats smartly.

5. Do not buy nondescript millinery. Make every hat tell a story, fit an ensemble, fill a fashion need. It should have distinction twenty feet away.

6. Yet, the fancier a costume is, the simpler the hat must be.

7. It is much better to buy a hat to go with one or two garments than to try to have one hat go with everything.

8. A hat is the frame for your face.

9. A hat must be in the right relationship for your face, head, and style of costume. Study the relationship to your face, head, and hair style, not only from the front but from the side and back.

10. Even the most perfect figure would look awkward in a wide hat combined with a boxy jacket and a wide skirt.

11. A short, heavy figure looks even shorter and becomes top-heavy under a large, wide hat.

12. If your shoulders are extremely wide and your face is full, a hat should not extend in width beyond the middle of your shoulders.

13. Never buy a hat until you have had a look at yourself in a full-length mirror.

14. When you are buying extreme fashion in a hat, particularly if it is very expensive, remember that you may only be able to wear it that one season.

15. The hat made from matching fabric of coat or suit can lack smartness; too often a touch of the fireside is very apparent. Too much sameness is seldom smart. If a matching fabric hat is your choice, make certain it enhances you and your costume.

16. A hat should never be darker than your shoes.

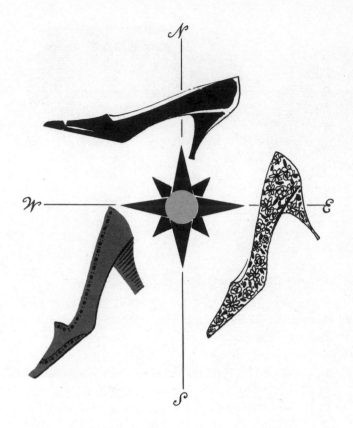

Shoes. 1. The style of the shoe is most important in relation to the shape of the leg and ankle. Pumps are universally our most flattering shoe style.

2. Pumps are also our most universally accepted shoe style with most costumes.

3. Select a height of heel that is flattering to your legs and ankles, and with which you can walk gracefully. To be well dressed, you should wear heels.

4. Don't wear bright-colored shoes or shoes with a fancy trim unless your feet are pretty.

5. Heelless and toeless shoes are difficult to wear.

6. Analyze the style of a shoe at the back in relation to your ankle.

7. Of plastic shoes, we say that unless you consider feet to be pretty, why display them? Occasionally, plastic slippers with an appropriate trim have been attractive for evening wear only.

8. Don't spoil an evening ensemble with your go-with-everything black suede pumps! Satin or silk shoes are a smarter choice. Satin pumps in different colors (they may be dyed to match a dress or made a full tone deeper) are a classic addition for after-five clothes.

9. If you choose patent leather pumps, don't wear them with sport clothes.

10. Patent leather shoes can be worn almost the year around, depending upon your costume, and they are perfectly correct for informal, after-dark occasions.

11. High-cut shoes and heavy ankle straps steal length from your legs. Ankle straps make your ankles appear wider.

12. White shoes belong only with a white or very light-colored costume. They are never the correct groundwork for a dark costume.

13. Nothing spoils an outfit more than timeworn shoes, and shoes which are obviously out of style.

14. Shoes covered to match a costume (print shoes for a print dress for example) have proved to be a smart choice. They may complement a costume more than any other shoes you can purchase. However, you should not repeat the print *any other place.*

Handbags. 1. Select a size of handbag in relation to the number of articles you like to carry in one, so that it can retain its shape.

2. The size of a handbag should also be chosen in proportion to the height and size of the wearer. Hang it on your arm and analyze its size for you, or study yourself in a long mirror.

3. If your gloves and bag match, and moreover if they match your costume, they will increase your height.

4. Anyone has a difficult time caring for a white bag.

5. Because you have a beautiful alligator bag, you needn't purchase alligator shoes to match. Ensembles are often overdone, and a beautiful alligator bag can very well stand on its own with leather pumps.

6. Huge leather carry-all or tote bags are wonderful for traveling or for shopping, but they should never be carried after five o'clock. When you leave home for a full day, you can put a small clutch bag in your tote bag.

7. Think twice about a bag that is decorated to look like a flower or fruit garden, or Sherwood Forest! One you can see through reveals everything you have inside, also!

8. A contrasting bag calls attention to your hips.

9. You will always look more graceful carrying a bag up on your arm at your waistline, instead of having it dangle at your side with your arm straight down.

Gloves. 1. Gloves are traditionally a sign of dignity, but not soiled ones! When wearing light-colored ones, always carry an extra pair in your purse.

2. There are informal occasions when gloves aren't needed, but always wear them on the street, even without a hat, to church, the theater, and luncheons.

3. Longer gloves are more dignified than short ones, although the length of gloves is partly controlled by the length of sleeves, by the attractiveness of your arms, and by the occasion to which they are being worn. Very formal occasions call for gloves extending up over the elbows.

4. Short, white gloves are smart with anything sleeveless or summery; or short black ones may be more interesting. Short, wrist-length gloves are often a smart choice for party dresses, also.

5. In some cases, no other accessory can do as much for a dress as the right pair of gloves.

6. Are white gloves always good? It is very often better to coordinate them with the colors in your outfit. Neutral shades are very popular.

7. If your forearms down to the wrist are overly heavy, avoid wrist-length gloves.

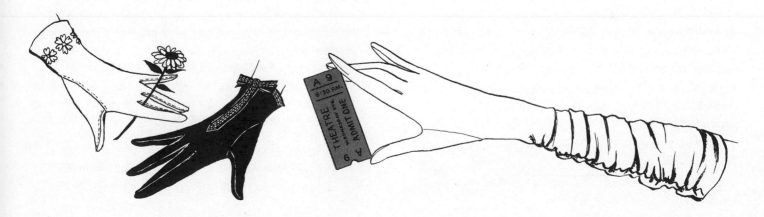

Jewelry. 1. Don't have everything match—those *sets* of earrings, bracelets, necklaces, etc!

2. Pearls are our most flattering jewelry and are correct at all times. Just be certain to get the right size for *you*, and the right length for the neckline of the dress.

3. Dare to be simple; if there are fancy buttons, let them be the focal point in your costume without other adornment. Don't obscure a beautiful neckline with a distracting necklace or beads.

4. Unless your pin expresses something about your costume, leave it in your dresser drawer, and look at it there! Wear pins imaginatively placed or paired—slipping out of a pocket or from under a lapel!

5. All pieces of jewelry worn at the same time should have some relationship to each other.

6. The rhinestone category in jewelry is still best after five, except that possibly a *small* rhinestone pin may be worn during the day.

7. Limit yourself on bracelets and rings if your hands aren't pretty, or they will reveal your age.

8. You should wear only one bracelet on plump arms, but you can wear many on long, thin arms.

9. The way your ears are formed will decide what kind of earrings you can wear.

10. If you have a round face, wear long pearls instead of chokers, or a pendant on a slim, long chain, and narrow, drop earrings instead of round or square ones.

11. If you have a long face, wear button or cluster earrings and a large choker high on the throat.

12. A thin neck is camouflaged with space-filling jewelry.

13. Buttons of a large ball variety add thickness to a thin figure.

14. If you are heavy, avoid glittery or eye-catching fastenings, large buttons, and massive pins.

15. A change of buttons or other trimmings on clothes you buy can make them more becoming and expensive looking.

16. Many women over 50 find silver jewelry more flattering than gold.

17. If you have an over-sized bust, avoid wearing too long or too bulky a necklace.

18. Avoid wearing too many pieces of jewelry at one time, such as a pin on a hat, earrings, necklace, lapel pin, watch, bracelets, and rings.

19. A touch of excitement can be added to a costume with the right jewelry: pearls, for example, with a strand or two of jade-green beads in their midst; emerald beads on a blue dress; or a huge turquoise pin on an herb-green dress.

Miscellaneous things. 1. Before selecting clothes, get the right foundation garments. You will need several styles for various styles in clothes. Try on a foundation before buying it, because all styles aren't for you.

2. You can wear too much underneath your clothes. A half-slip is better than a full slip in many cases.

3. Your slip should harmonize with the color of your costume. Lingerie is now available in many colors. Many people will sew a band of matching color of dress fabric around the lower edge of their slip. Always wear black lingerie with a black costume.

4. Do not go stockingless! Seams and long heels improve legs and ankles.

5. If your legs are ungainly, keep the color of your hose inconspicuous. However, darker hose are more flattering to legs than lighter ones. Remember that very dark hose look hot.

6. Many people who are all dressed up will pull bathroom tissue out of their handbag! You should always carry two handkerchiefs, and they should harmonize with your costume.

7. Many times for a dash of color or design, you can wear a scarf for a belt or hang one over your regular belt.

8. For the most part, best designers say that artificial flowers should be kept on hats! Exceptions are those made of self-fabric or lace.

9. A suit worn without a blouse, or with a blouse or little sweater the same or harmonizing color of the suit, gives the look of a complete ensemble. The little white blouse worn with every suit seldom gives the look of a complete, smart ensemble.

10. Short, stout people should wear short-haired, flat skins, such as broadtail and caracul; never bushy, bulky, or long-haired furs.

11. The style of a stole is as important as the style of your clothes. A full-length or medium deep stole with short ends is best on most people.

12. If a stole has long ends that come below the waistline, the individual must know how to wear it smartly.

13. A person who is short or wide should not wear a stole with long ends. Analyze the length of the back, also, for the size and height of the person. Neither a stole nor a jacket should cut a short person in half.

14. It is important that the cut and the shape of the collar, as well as the amount and the hang of the fullness of the stole, be right for the individual.

15. The color of the fur you choose should be just as flattering to your hair, skin, and eyes as any other color you choose for your costumes.

16. A little jacket or stole is often needed to cover that too bare look.

17. Thin, plastic raincoats are not a "well-dressed" choice.

18. Glasses should be a part of your personality, and not of your fashion wardrobe. Unless you have an extra, novelty pair for sportswear, the shape and color of the frames should blend with your facial tones and hair coloring.

19. If you wear glasses, especially if they have detail of trim in the frames, you *must* limit the amount of detail at your face in earrings, necklaces, and veiling or trims on a hat!

20. Select buttons under a good light. Use contrasting colored buttons with care so that they don't look spotty. When used, these buttons are generally the trimming detail of the dress; so keep the dress simple in design.

Choice of color

Fortunately or unfortunately, color is our greatest accessory; it can destroy beauty as well as create it. Here are some tips on color to help you in developing the art of being well dressed.

1. No one should wear colors she does not like.

2. Give much thought to what a color will do for your figure, your skin, your hair, and your eyes.

3. You should think of color combined with texture, not of one or the other independently. Texture affects color—turquoise in satin versus jersey, red in soft velvet versus hard-finished taffeta, and brown in harsh gabardine versus soft fleece.

4. Take it easy on bright colors—a touch of gaiety will do a trick, but an avalanche will bury you.

5. Only those who are vivid in coloring should wear vivid colors in large areas. The less vivid the individual, the less apt she is to transcend strong color. People with soft coloring need more subdued color to enhance their own.

6. Sometimes, a little tint in the hair makes all the difference in the colors one can wear. It may only be a tint that is sprayed on the hair.

7. Lighter colors should lead to the top of the costume.

8. Remember that the eye always goes where contrast takes it.

9. Consider whether a color will soil easily or show lint, and how much care will be necessary to keep it looking fresh for the places it will be worn.

10. Never wear more than three outstanding colors in a costume at one time (unless a print shows more);

never use the same color more than three times. For example, a white hat and white jewelry would be desirable with a black dress, handbag, and shoes. Matching black gloves would be counted with the handbag. A bright hat, bag, shoes, gloves, and jacket would never be good; a bright hat and belt would be smarter.

This rule serves as a wonderful guide, but there can be exceptions to it; for example, some women can wear all black or all beige costumes smartly.

Then, the monochromatic color scheme is effective and popular in today's fashion. Shades from light blue to navy or from pale beige to dark brown are good examples of this approach.

11. Decide on basic colors in your costume planning; keep them firmly in mind, even if you are only shopping for handkerchiefs.

12. Fine fabrics cost money; color does not.

13. Overweight women will always look slimmer in colors medium to dark in tone. Dull black and subdued or grayed colors have a receding effect.

14. Since dark values of color tend to make a large figure recede, the opposite in light, warm colors should be used by the tall, thin figure.

15. Black has everything—dignity, elegance, sophistication, simplicity, chic.

16. Black and white are always considered dramatic together and can never be beaten as a color combination. Almost everyone can wear black and white. However, the aging, with graying hair and fading color, should beware of black, since it absorbs light and tends to drain light from the face.

17. Light, neutral hair is reduced to further drabness with beiges and tans.

18. Neutral colors, such as light beiges and grays, tend to be less interesting at night. Dominant colors, as well as black and white, are fine after dark.

19. Pink is considered a universally becoming color.

20. Turquoise is a good color for most women.

21. Deep pink and turquoise are good night and day colors.

22. Other delightful evening colors are gold and gray-blue.

23. Pale blue is a wonderful color.

24. Bravery never goes out of fashion in combining colors—electric blue with masses of white pearls, shades of blue with shades of green, brilliant plaid

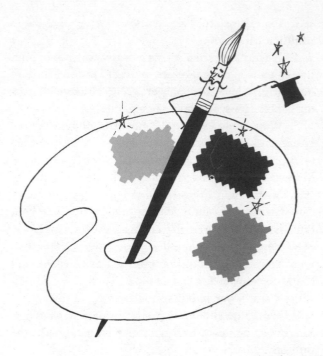

with chunky gold, or creamy salmon with turquoise jewelry.

25. Instead of thinking that you cannot wear a certain color, learn that you cannot wear certain shades of the color. You can wear almost any color if you find the right shade or value that is softened or grayed enough for you.

26. Do not wear two grayed colors at one time.

27. Some women can wear navy much better than black for a basic color in their wardrobe.

28. Always ask yourself what impression you wish your scheme to convey—vitality, drama, joy, dignity, maturity, or conservatism?

29. Lastly, the total effect of any color harmony depends not only on the colors chosen but also on the proportion in which each occurs and where they are placed with relation to each other.

COLOR IS PURE MAGIC!

Necessary care of clothes for being well dressed

Here are some points to keep in mind:

1. Keep hemlines even; it is a full-time job, but very important.

2. Dress and skirt hems should be ½ inch shorter than full-length coats.

3. Your slip should never show. Always check in a full-length mirror.

4. Never hang up a garment wrong side out. Place it on a hanger grain perfect, fasten some of the buttons, and close zippers. Never leave belts on dress belt loops. Hang them separately.

5. Wire hangers are only meant to be carriers between the cleaner's and your home. Clothes must be supported by stronger hangers to hold their shape. Make certain skirts do not dip in the center with the style of skirt hanger you choose.

6. Do not wear limp veils, unpressed bows, or bent flowers on hats. Press hat veiling between two layers of wax paper to freshen it.

7. Your shoes should be polished or brushed, and the heels kept neat at all times. Don't forget to polish or brush your handbags, also.

8. Handbags should be pretty inside as well as out. Keep your compact fresh and clean, and replace the puff often.

9. Do not wear any garment that has a spot on it or is soiled in any way.

10. Discard old, limp gloves that you haven't worn in a year. Perfection in each pair of gloves is more important than the number of pairs you own.

11. Keep coats and other outer garments separate from lightweight clothes in closets, so they won't push them out of line.

12. If you wear a panty girdle, you can tuck your blouse inside it. This keeps your blouse in place with skirts or shorts.

13. Aside from carrying an extra handkerchief and pair of gloves in your handbag or tote bag, there will be occasions when you will want to carry an extra pair of stockings.

14. For perfection in grooming, clothes must be immaculately clean and pressed, with buttons, hooks, and snaps securely sewn in place, shoes well polished, heels neat, gloves well-fitting, and stockings never loose or twisted. Work toward a head-to-toe fastidiousness of grooming and dress.

15. To keep stocking seams perfectly straight, fasten your back supporter in the exact center of the seam at the full fashion mark (unless, of course, your girdle does not have a supporter placed at the back).

16. Never wear jewelry with missing stones.

17. Make-up and perfume will discolor some kinds of jewelry if you are not careful to prevent them from getting on it.

The perfect dress, suit, or coat

The lines of your costume

The lines of your costume produce movement, causing the eye to move in different directions—vertical, horizontal, and diagonal! Study every line of a pattern or dress in relation to your figure before you select it. Remember always that simple lines are best on everyone.

When studying the following rules about line, keep in mind that there will always be exceptions to every one of them.

Lines of an entire ensemble. 1. The favored standard figure is approximately 5 feet 6 inches tall. If you are taller, select lines to make you look shorter. If you are shorter, select lines to make you look taller. One of the most important factors of the ideal woman's figure is that the width of the shoulders and hips as measured across the front are the same.

2. Always wear a straight skirt with boxy garments that end below the waistline.

3. Never end a jacket or other garment at the widest part of large hips, or do anything else that will create a horizontal line there. Keep the jacket shorter or longer to form a pleasing proportion for the figure.

4. A semi-fitted jacket is recommended for the person with large hips and large bust. A jacket that is not fitted too close to the figure is younger and more becoming to most women.

5. If you are short and wish to look taller, jackets should match your skirt or dress, and short or long, keep them slim. A one-piece dress with a jacket is better than skirt, blouse, and jacket.

6. Double-breasted lines are good on the tall, slim figure, while single-breasted lines are better for the heavy or short figure.

7. If you or your daughter are chubby, do have vertical lines in your costume, and not horizontal lines. Upward slanting diagonals sometimes give added grace. Emphasis should always be within your silhouette.

8. If you are chubby, coats should be full length preferably, either straight or semi-fitted, with easy, long-fitting sleeves. The bulk of a big, loose coat with too much detail makes you appear larger.

9. A cardigan, chosen in the right length, is universally becoming with a straight skirt.

10. Many styles of dresses that button down the front make you look older and heavier.

11. A sheath dress does not look well on a heavy figure—a skirt with some fullness is more becoming.

12. It takes a tall, slim figure to wear a dress with a wide look in the cut of the bodice and in the cut of the skirt. Many figures can wear a wide look one place, but not both.

13. If you are short, wear slim, one-piece clothes; do not break the line with color or wide belts. Wide belts make you look squarer.

14. A ½-inch wide belt (narrow and inconspicuous, of self fabric) is more flattering to the figure with a thick waist than a one-inch belt. Even a better choice may be to finish the waistline seam with a piece of bias (similar to middle photo, page 194, *The Bishop Method of Clothing Construction*), and eliminate a belt.

15. Contrast at the waistline is good for the tall, thin figure, since it will draw attention to this area and thereby cut the apparent height.

16. Much shirring on a garment makes the figure look heavier.

17. The tall, thin figure needs lines which attract the eye in a horizontal movement. She should avoid the clinging, too tight silhouette, and is best in the flattering effect of unpressed pleats, tucks, and other forms of flowing fullness. Boxy jackets are better than figure-revealing, fitted ones. In length, they may come to the hips, or any length below the hips which gives a pleasing proportion. Rarely are bolero jackets good for this figure, because they give the effect of too great length of limb.

18. Short shorts are for the teen-ager; bermuda shorts are more attractive on most women. Long pants are the kindest to figures with unattractive legs but good hips. In any length, slacks are flattering only to the women who have slim hips and a flat stomach.

19. Capes are good on the tall, thin person, since they tend to reduce height.

20. Do you enjoy the real pleasure of fashion to the extent that you can and will wear a coat smartly that has to be clutched all the time? All fashion requires the right wearer.

The lines of the bodice. 1. A strapless dress is a poor choice for the short person or one with thick shoulders or a large bust.

2. Wear sleeveless dresses only if your arms are slender, and never if you have overdeveloped arm muscles, any sign of flabbiness, or a heavy upper arm. Of course, you should not wear sleeveless dresses if your arms are too thin.

3. Dolman and raglan sleeves are good for square shoulders. They are not recommended for the figure that has narrow or sloping shoulders and large bust and hips. Set-in sleeves minimize the bust and hips.

4. Longer sleeves tend to camouflage the size of large forearms. The most flattering sleeve length is below the elbow.

5. Wide, loose sleeves tend to widen the figure.

6. Long, thin arms are concealed by the dolman or by wide, full sleeves with a band or deep cuff.

7. Consider carefully contrasting cuffs or puff sleeves; they make the figure look heavier and wider.

8. If the color of a sleeve is in sharp contrast to the color of your skin (black versus beige, for example), then the sleeve length is even more important to consider for your arm.

9. The all-in-one sleeve is simple to fit, almost universally becoming (see point 3), and can't be topped for comfort.

10. High round necklines are a focal spot for jewelry.

11. A square neckline fits better if it is slightly curved at the corners.

12. Necklines should be cut close at sides and back for a stout figure. The slightly lower neckline in front, such as a narrow square, will cut the thick look of shoulders and bust.

13. Collars for a heavy figure should be narrow and relatively flat. The shawl collar with a V neckline is one of the best. Outstanding lapels attract the eye and should be scaled to balance hip width.

14. The thin figure needs cover-up necklines, such as a cowl collar, scarfs, and soft bows. Very becoming are rolled or flared collars with curved edges that add thickness and width to thin necks.

15. The size of a collar must be scaled to the size of the wearer, and the shape must be flattering to the shape of the wearer's face and neck.

16. Yokes all the way across the front or back of the garment will widen the figure, and tend to emphasize the bustline or round shoulders.

17. A soft rolling collar, generous flat collar, bolero, cape, or short, straight jacket is flattering to the figure having round shoulders.

18. Bulky versions of pockets and bows will cover hollow chests.

19. Draped, bias necklines are not good for the figure with a large bust.

20. A figure with sloping or narrow shoulders may be built out with shoulder shapes. A wide, high lapel also helps to create the illusion of a straighter line. As we have learned, this figure is not good in dolman sleeves.

21. With a large bustline wear a bodice with slightly draped fullness or ease, rather than one which is revealingly smooth and tight. Such a bustline may also be broken by some vertical or diagonal movement.

22. A rounded, low collar is not good for a large, low bust. Choose a wide collar at the shoulder line.

23. Easy, bloused waistlines are good for the tall, thin figure.

The lines of your skirt. 1. The length of your skirt has a great deal to do with the appearance of your legs. When skirts are short, they will be more flattering if they end as nearly as possible to the widest part of your back leg muscle; the legs will then taper gracefully from hemline to ankles. This point occurs at different heights on different people.

2. Don't lose chic by having a full skirt too long. A flared, pleated, or gathered skirt may be shorter than the straight skirt.

3. Straight or moderately wide skirts are much more slimming than wide ones.

4. Three or four gores in the front and in the back are better in proportion for a large figure that enlarges when seated, rather than one or two gores.

5. Two gores in the back are more becoming for the figure that can wear such a slim skirt than one piece in the back. The skirt will also hang and wear better. You may add a center seam when cutting the garment, if the pattern is cut all in one.

6. A wide skirt will make a small waist look even smaller.

7. If you are tall, you will look well in the widening look of unpressed pleats. If you are short, they are not for you unless they are modified.

8. Tubular, bouffant silhouettes in skirts when softened with pleats, peplums, tiers, or tunics give an effect of flowing grace for the tall, thin girl.

9. If your hips are too large, do emphasize your shoulders and neckline. Do not use hip decorations, such as patch pockets on both sides of the skirt. Skirts should be simple, easy, soft in line, and never tight or narrow.

10. For the smaller and more youthful figure with large hips, an equally pleasing effect may be produced by keeping the bodice and waist trim, and covering up the heaviness of the hip with a modified full skirt.

11. A prominent abdomen can be minimized with slight fullness in the skirt on either side of the protrusion, or with a finger-tip length jacket.

12. The bustle silhouette returns every so often. It is always a limited fashion, because few people are tall enough and thin enough to wear it well.

13. Bias-cut skirts are figure-revealing, and are difficult for the majority to wear smartly. They are also a problem to keep straight at the hemline in most fabrics.

24. Soft tucks or gathers are preferred at the waistline of the bodice for the figure that has a roll above the waistline or a large rib cage.

25. A straight-lined sheath will also accent the above disturbing line at the waistline.

26. Today's fashion calls for a normal shoulder line; therefore, if you do not have normal shoulders, use shoulder shapes (pp. 158–159, *The Bishop Method of Clothing Construction*). If only one shoulder is out of line, just one shoulder shape will be needed.

27. A curved line will add roundness and weight where it falls on the figure. If the line is directly over the bust, it gives an illusion of a full bust.

28. Points are slenderizing to the figure, for they lead the eye up and down in the direction of the point. The edge of a sleeve that is shaped into a point will slenderize a heavy arm. A long V at a neckline is flattering to a short neck and round face. Points which extend into the skirt decrease the width of the hips. A point extending from the shoulder in the direction of the waist increases the apparent width of the shoulder and minimizes the waist.

29. A beautiful bodice to a dress must have definition of fit and must also be in perfect line for the individual figure.

14. More women look better with released tucks in the front of a skirt instead of stitched darts. Where they are placed depends on the individual figure.

15. The standard length of darts on patterns and ready-to-wear will frequently need to be adjusted for most individuals. To be most becoming, they should end short of the fullest part of the figure, such as the bust and hips.

16. The length of pleats must be in correct proportion to the height of the individual figure.

17. Pockets of bulky versions will cover bony hips.

18. Overblouses are widening to your hips, especially when in contrast.

19. Square corners on the lower edge of a jacket will be more slenderizing to your hips than rounded or slanted ones.

20. If you have wide hips, never narrow your shoulders. This makes your hips seem awkwardly large by comparison. If your hips are exceedingly wide, use sharp contrast only at the top of your figure—in your hat, at your neckline, across your shoulders—and wear dark colors with simple lines.

21. A short overskirt or a side drape decreases the apparent height of the wearer and adds width to the skirt.

22. The width of a waistband on a skirt is determined by the set of the figure at the waistline and by the height of the individual.

23. A beautiful skirt to a dress must have definition of fit and be in perfect line for the individual figure.

24. Nevertheless, in order to look well dressed, you must be comfortable in your clothes and look as if you are able to move inside them. Beauty today comes from simplicity and comfort in dress.

The fabric you select

Fabric must be chosen to be right for the time of day and year, for the occasions for which it will be worn, for the figure, and for the pattern to be used. Color has been discussed in Chapter 1, "The Well-Dressed Woman."

1. Examples of daytime fabrics are flannel, wool crepe, tweed, silk linen, wool jersey, cotton, linen, and miracle fabrics.

2. Examples of fabrics for traditional occasions (big evenings and weddings) are satins, lace, organdy, brocade, velvet, and chiffon. Lace, chiffon, peau de soie, and taffeta can be worn the year around. Particularly appropriate for fall and winter are velvet, satin, and brocades; and for spring and summer, thin prints, organza, and organdy. Be careful about your fabric choices for evening. A solid color, perhaps crepe or faille, stands out in a group, and can be worn more often than something with much pattern and design. The lines should be simple and never cluttered.

3. Never use glittery, metallic fabrics for daytime wear.

4. Silk, linen, and wool are marvelous, rich fabrics: silk drapes well; linen is strong and durable; and wool is conforming and enduring.

5. Jersey is a wonderful fabric, but because of its characteristics, the pattern should be chosen carefully (see page 97, *The Bishop Method of Clothing Construction*). If the pattern is not chosen wisely, jersey will cling and be figure-revealing. Suit-weight wool jersey has much character. If you are planning a trip, add jersey to your selection. It will pack easily, travel well, go anywhere, and wear a long time.

6. Many cottons require too much pressing.

7. Thick, bulky, or shiny fabrics are not good on a large, heavy figure. Not only will shiny fabrics increase the apparent size of the figure, but they will reveal all the contours of the body and accentuate the lines in the face. Velvets and velveteens increase size because of their pile surface. Dull, broken-surfaced fabrics, such as heavy crepes, will absorb the light and reduce apparent size. Bulky, napped, and pile fabrics will add inches through their thickness and rounded folds. The large, heavy figure should never wear stiff, crisp, wiry, sheer, clinging, or flimsy fabrics, since they increase the size of the wearer. Good choices are lightweight tweeds, sheer woolens, heavy silk, wool crepes, semisheers, and linen.

8. A tall, thin girl needs napped fabric and soft fabrics that drape.

9. Transparent fabrics are figure-revealing unless used in several thicknesses. They are never attractive with buttons, buttonholes, and facings.

10. A hard finished, worsted fabric is not a good choice for an older person.

11. Harsh, firm, unyielding fabric is not good on the figure that is difficult to fit.

12. Stiff fabrics, such as taffeta or faille, will not yield for set-in sleeves; raglan or cut-on sleeves will continue in fashion for this reason alone. Stiff fabrics, unlike soft ones, do not lend themselves to any straight, closely fitted line.

13. Fine fabrics, such as silk, should never have a lot of cutting and stitching done on them with intricate designs in a pattern.

14. Surface fabric with an interesting texture should never be chopped up.

15. Some fabrics can be combined beautifully, such as a gray flannel suit with a white satin vest or blouse, a wool crepe suit with a silk print lining and blouse, or a printed silk dress with a wool crepe jacket. A coat of a tweed fabric should never be worn over a satin dress.

16. Many fabrics handle well in soft, heavy folds. Often a fabric such as voile would have little texture interest when used on flat surfaces, but it gains distinction when its surfaces are broken into tucks and folds. Chiffon and georgette hang in a similar way to voile.

17. Heavier satins, taffetas, and failles take on a look of elegance.

18. Wool, silk, and velvets lend themselves to small folds in a soft, clinging fashion.

19. Remember always the importance of selecting fabric carefully. It must have the right weight or body for good results in the job to be done.

20. Analyze fabric by the yard or in a dress by holding it up to your face before buying it.

21. A black or navy dinner dress in a lightweight wool or in silk crepe is a must in everyone's wardrobe.

If you choose a design

We are discussing here the choice of prints, polka dots, stripes, and plaids for the individual. Once chosen, what consideration should be given to the way they are made up, and where should they be worn?

1. When making up a garment from design, the structural lines must be kept simple. Emphasize the importance of the design. You liked the fabric when you bought it; so the design should be kept in first place, and the lines kept simple. Many structural lines of a dress are lost in the design.

2. Study the design of your fabric and see if it will tell you what to do. The design of the dress should harmonize with the design of the fabric, and it may even suggest its own trim.

3. Reversible prints, brown on white in a dress, white on brown in a jacket, are popular in today's fashion.

4. Two opposing designs should never be worn together, such as a plaid skirt with a print blouse.

5. You will tire quickly of bold prints and plaids or large checks and stripes. Ascertain if it will be possible to match the design of the plaid in the lines of the pattern you select (see pages 96–97, *The Bishop Method of Clothing Construction*).

6. If you are tall and thin, you can wear a large blanket plaid, but if you are short and stout, avoid it.

7. Bold stripes and polka dots are considered *hard* in appearance for many people to wear. Perhaps, using them for a trim on a garment (overcollar and cuffs on a suit) would be a smarter choice than using them for the entire garment. Yet, while coin dots may seem to be too much, polka dots, scattered or small, are easy to wear. Polka dots unevenly spaced are usually more interesting in design.

8. Choose vertical stripes if you feel that horizontal stripes make you look wide.

9. On a heavy figure, prints should be small in design against a darker background. Polka dots should be very small. Prints that lead the eye in a circular or swirling movement suggest rotundity. Prints that are widely spaced and in strong contrast attract attention to size.

10. With a two-color tweed or check (such as black and white), the garment generally becomes more interesting when both colors are brought out separately in the costume as sharp contrasts. An example is a white collar with a black binding. However, the sharp contrasts may be done with accessories.

11. With a bright design, any contrast would be a dark color in the fabric to tone it down, such as a black velvet bow or belt. With a subdued design, any contrast would be a bright color to pick it up, such as yellow linen lapels on a jacket.

12. Avoid straight seams in a check or stripe when selecting your pattern. With red and white check, for example, a straight seam may bring together two red or two white lines; this will not be pleasing.

13. We must learn to recognize good prints, to select them to be right for the occasion, and to have them express the personality of the wearer. Many women are afraid to wear prints; they feel that prints compete with their faces, figures, and personality. They are surely right unless they know how to select the proper ones.

14. Women very often look cooler in solid colors than in prints.

15. A few well-dressed women rarely wear prints, while others have them in their wardrobe the year around. Whatever is your choice, the following facts must be recognized.

a. Strictly realistic treatment of animal and plant forms and of scenic landscapes are never suitable for wearing apparel. Good prints may have their motifs based on nature, but the designer will have added something creative of his own. In other words, choose an artist's print rather than a gardener's.

b. The shapes or motifs should be interesting in contour, and so arranged as to make a pleasing rhythm.

c. The spaces which form the background areas must have interest in themselves so that the fabric as a whole will be pleasing.

d. Do not purchase a print that lacks any particular significance.

e. Prints suggest different things to us. Some are so exotic, bright, and unusual as to be suitable only for gala social evenings, vacations, or cruise wear. Others give the impression they belong to youth and are gay and amusing. Many seem dainty, refined, ladylike, as though they were intended for delicate, feminine women. Certain allover prints give the impression of force, vigor, and drama, suggesting their use by sophisticated, dignified women. Finally, there are those that are conventional and conservative, admirably suggesting the conservatism needed by the mature woman or one in business.

Alterations of ready-to-wear

We have had countless requests to do this chapter in our second book, because surprisingly few people know how clothes should fit. In putting together this chapter, we have tried to show the most common alterations necessary for the majority of people.

Whether you do the alterations yourself or have someone do them for you, it is important to know how clothes should fit! The average individual wants things much too tight, while others want them too loose. A garment that is too tight will accent every figure fault and will make your figure appear larger. You should always conceal rather than reveal your figure defects. Furthermore, there must never be any strain on the crosswise or lengthwise, or the garment will never look or feel comfortable.

Even the most expensive clothes can never have a quality look, unless they fit well. Alterations are not difficult, if you just know how they should be done.

Clothes are manufactured in many different types and sizes—Junior, Misses', Women's, Half-size, Diminuettes, Petites, etc. Learn what is best for your figure. A common alteration is to change a regular size into a half-size, but you should avoid it, if you can.

In addition, you should remember that the style of the garment governs what alterations can be made. In other words, you must consider whether the alterations will impair the style of the garment. Or, stated in another way, you must understand the silhouette of the garment to judge the alterations wisely.

Then, too, the lines and details must be right for the proportion of the individual figure. Many people buy clothes because they like them, and give no thought to whether they can or cannot be fitted to their figure.

Clothes that fit properly are comfortable. You can forget yourself in them, be self-assured, and at your best. Compared to the past, comfort and freedom are highlights of today's fashion.

The right style of foundation garments and a proper fit for the individual figure are extremely important in helping you to determine what alterations will be necessary.

When you have made a decision to purchase a garment in a store, the fitter is called. Then, just as the fitter has the prerogative to turn down a sale if she thinks the garment cannot be altered properly, so do you, if you feel the fitting will be too involved after the suggestions are made. To make a garment look fashion right is a fitter's business, but it is also your personal concern. Too many alterations are always questionable.

However, more women are learning to do their alterations at home, because the cost of alterations continues to increase. For example, in one large department store, out of 49 garment sales, 10 are pin-fittings to be done elsewhere. This same store had 55 fitters ten years ago, 30 four years ago, and 17 today. The majority of the stores report between 40% and 50% of their transactions are pin-fittings. Yet, some of the alterations have been minimized in recent years because of the many types of clothes on the market. The consensus of opinion among alteration departments is that faulty manufacturing very materially affects alteration costs. Sometimes, garments are even sent back to manufacturers because the cut is wrong. If the garment is cut off grain, it can never fit anyone nor ever be right in line. Thus the wise shopper will always be grain conscious.

The most common and routine alterations are at the waistline, side seams, and hems. Next in line are adjusting shoulders and correcting sleeve lengths. A garment that fits the upper part of the body is the best buy; skirt alterations are easy. Shoulder and neckline alterations are more difficult and are apt to be less satisfactory than skirt alterations. They are also more expensive than skirt alterations.

In altering skirts, fitters often find that one hip is very unlike the other. It is usually easier and safer to take in a garment than to let it out. Seams that must be let out may present problems. The matching notches may be cut into them, the seams may be too narrow, they may show stitch marks, and they may have shine on the wrong side that cannot be removed.

Bust darts are lowered instead of raised 95% of the time in altering ready-to-wear. Sloped ones are better than straight ones on the majority.

Hem alterations usually cost from $2.00 to $7.00; adjusting sleeve lengths, from $2.00 to $4.00. Naturally, alteration prices vary with the section of the country. A machine hem in an alteration department costs less than a hand hem. It costs extra when there is an underlining in a dress. There is a machine that cuts the hem the correct width, and sews on the tape at the same time. Always be aware that hems cannot be let down if the underneath corners are cut away. Then, too, a full skirt will always appear to be longer than a slim one of exactly the same length, since the latter tends to ride up as you walk. It is interesting to note that one large department store altered 22,787 garments in 1960. Of these, 85.6% had hem alterations, but may have had other alterations, also. One-third of these hems were done by hand, and two-thirds by machine. That represented 25% of their business transactions.

We did not include anything on finishing hems in this chapter, because once a ready-made garment would be marked for shortening or lengthening, you would choose the hem finish most desirable for the fabric, as presented on pages 46, 82, and 142–143, *The Bishop Method of Clothing Construction.**

Almost no one is a perfect size. The chances that any ready-made garment will fit many women perfectly are very slim indeed.

All the suggestions offered in Chapter 7, *BMCC*, on "Fitting and Cutting-to-Fit" should help in understanding this chapter. However, all of them cannot be used in altering ready-to-wear, because you are very limited in what can be done after the garment is cut out.

We would like to call your particular attention to two sections on pages 47–48, *BMCC*, describing what is meant by a perfect-fitting garment, and then how a garment is handled when tried on for a fitting.

*Through the remainder of this chapter, this text will be referred to as *BMCC*.

1. The shoulder seam should lie on top of the figure—as if to cut the figure in half. That placement will be one inch behind the lobe of the ear.

Since the shoulder seam comes too far to the front on this dress, the back seam can be taken in, and the front seam let out as much as possible. For perfect alignment, the dress in the photograph looks as if it will need ¾ inch let out on the front, but the seam allowance may not give that much.

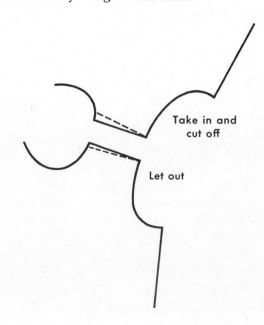

Take in and cut off

Let out

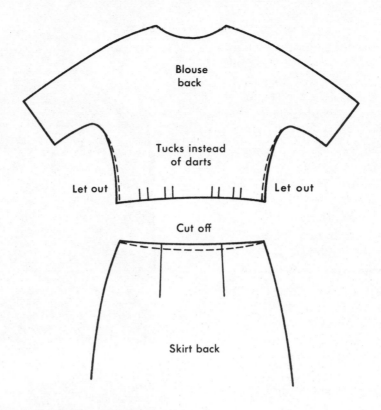

Blouse back

Tucks instead of darts

Let out   Let out

Cut off

Skirt back

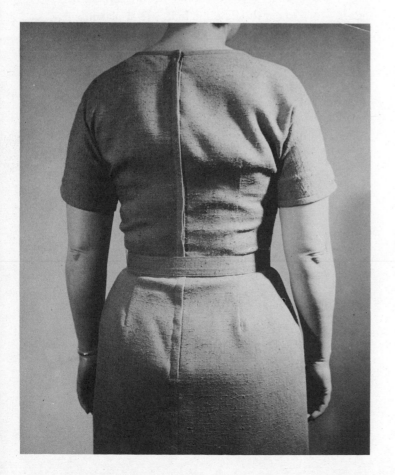

2. This figure needs more width in the back of the bodice. The seam allowance can be let out gradually at the sides of the back, and the darts changed to tucks to give additional ease. The softer bloused effect with tucks will be more becoming for this figure.

To eliminate the wrinkle at the center back of the skirt below the belt, the skirt can be cut out at the center back.

In making a dress for this figure, the skirt would need more width and height at the side back as indicated with sketch 6, page 67, and sketch 10, page 69, *BMCC*.

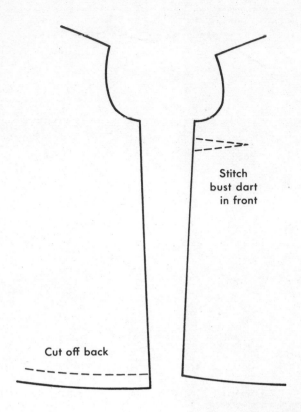

Cut off back

Stitch
bust dart
in front

3. When the side seam swings toward the front on a coat and the hemline drops at the back, the problem will be corrected by adding a bust dart. It should be made in line with the crown of the bust, and the same amount taken up in the dart must be cut off the coat back before the side seams are re-stitched. This coat looks as if it swings toward the front about two inches.

The pleat would be better if it lapped right over left because it would always have less strain when sliding under the wheel of a car, etc.

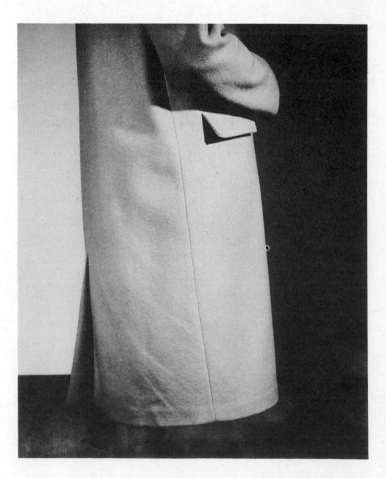

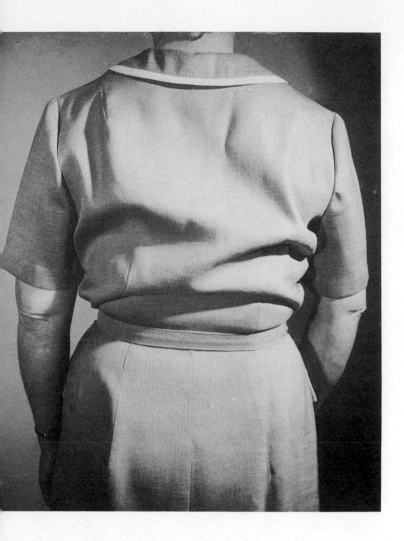

4. There may even be too much to alter in the back of this bodice, but what it needs is excess fabric cut off all the way across the bottom. A bust dart can be made in the front of the bodice, and if necessary, a small amount can be trimmed off the bottom of the front so that the side seams of the bodice will match with perfection.

This arm would look smarter if the sleeve were one or one and one-half inches longer.

The darts are too deep in the back of the skirt for this figure. They should be decreased and shortened, and the difference cut off the side seams of the skirt back. This is a common alteration.

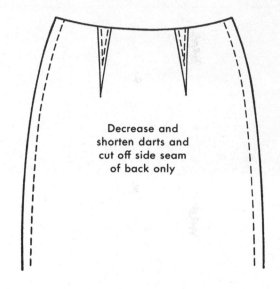

Decrease and shorten darts and cut off side seam of back only

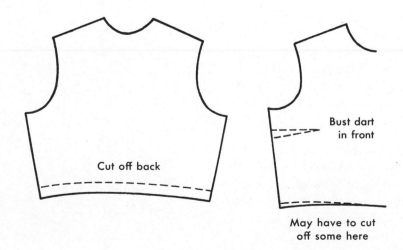

Cut off back

Bust dart in front

May have to cut off some here

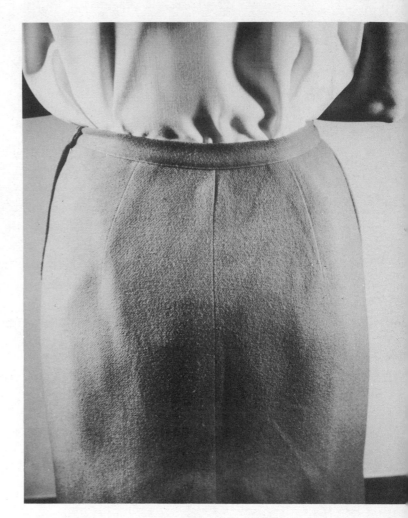

6. The curved hip darts accentuate the hip line, so this figure would be improved greatly if the darts were made straighter and two shorter ones were used instead of the one long and deep dart.

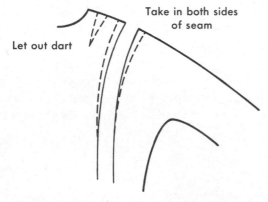

Let out dart

Take in both sides of seam

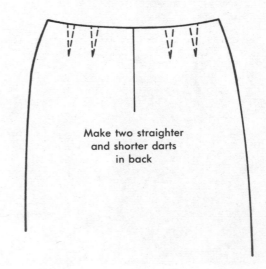

Make two straighter and shorter darts in back

5. The back shoulder dart that is showing below the collar should be taken out, because it is pulling the back off grain. Then, both sides of the seam should be taken in. We have many cut-on sleeves in the fashion picture, but they can present as many fitting problems as set-in sleeves.

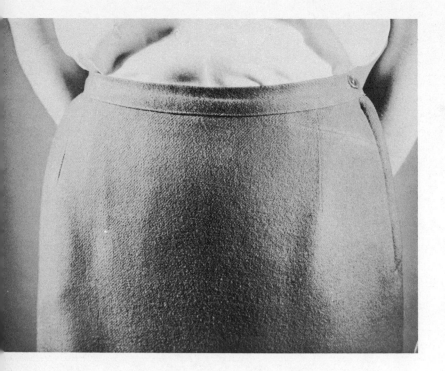

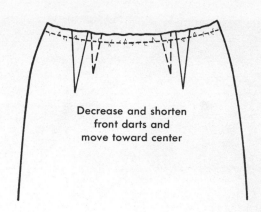

Decrease and shorten
front darts and
move toward center

7. The darts on the front of this skirt are too far apart and give the figure a wide look. The skirt was also too straight when it was sewn on the band.

To correct these, decrease the darts and move them closer to the center. Ease the difference at the waistline when stitching the skirt to the band.

Tucks or folds could be used on the front of this skirt instead of the darts.

8. This is a rather typical look in ready-to-wear. The pocket pulls apart because the skirt is too tight in the hips. It will not help to decrease the darts in the skirt front. Normally, the seams of the pocket and skirt are trimmed away, so it would not be possible to let the skirt out at the pocket edges. It is not wise to purchase shorts, slacks, dresses, or skirts that are tight in the hips and have pockets in the seams, because nothing can be done to correct the problem.

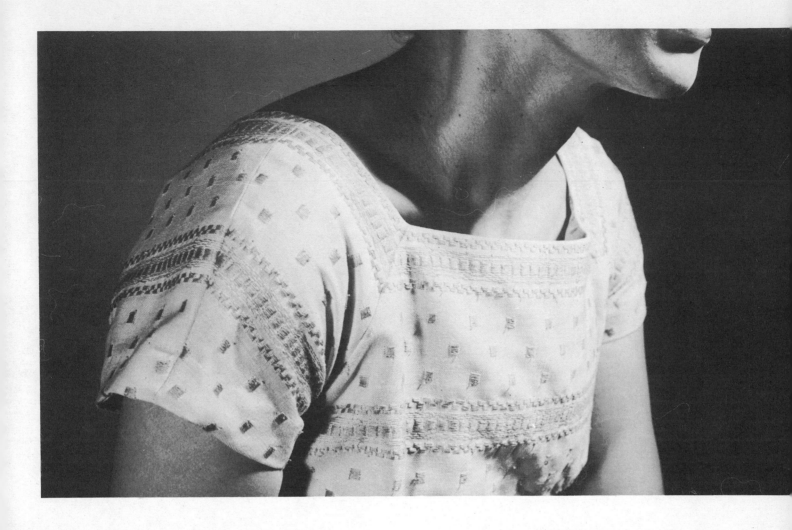

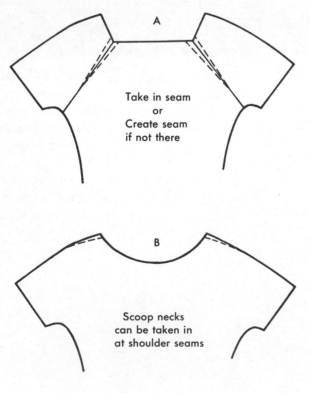

A

Take in seam
or
Create seam
if not there

B

Scoop necks
can be taken in
at shoulder seams

9. If a low neckline is loose or wide in the front, several things can be done to help correct it, depending upon the amount of the problem and style of the dress. In sketch A, the seams of the dress could be taken in deeper. In altering ready-to-wear, this seam is sometimes created in front or back or both places if it isn't already there in the style of the dress.

Sketch B indicates that larger necklines can also be taken in at the shoulder seams. This may have to be done equally all the way out on the seam. Sometimes the facing is taken off, and the neckline is eased. Then, the facing is made smaller before it is put back on the dress.

10. The bodice darts (A) from the waistline are too high; at no time should they come up over the crown of the bust. When the figure is low busted, it is better to change the bodice darts into two tucks.

This figure does not have enough bust for the depth of the bust darts (B); so they should be decreased, and the bodice shortened all the way across the front. Decreasing the dart will also help the side seam of the front fit the side seam of the back.

The skirt front is tight looking and needs tucks or folds instead of darts. Lifting the skirt front (a result of shortening the bodice, as the diagram shows) will straighten the tabs.

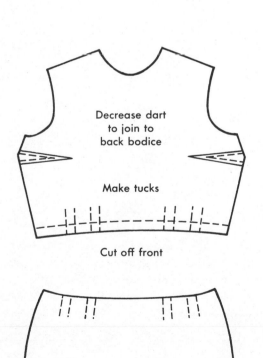

Decrease dart
to join to
back bodice

Make tucks

Cut off front

Make tucks instead
of darts

11. The sleeve is wrinkling because it needs to come forward in the armhole. The pin at A shows the amount the sleeve should be moved to the shoulder seam at B to straighten the grain across the sleeve cap.

The front shoulder is too wide and is falling off the shoulder bone. However, there is no extra width on the back, as the next photograph clearly indicates. Trim off the excess fabric as the sketch illustrates. The back shoulder seam will be eased to the front. When the side seam of the back of the jacket is cut down, as the next sketch indicates, it will be necessary to lessen the front darts.

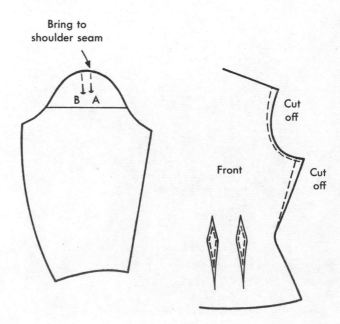

12. This figure is short-waisted. The pins at A identify the deepest part of the curved seams for the waistline, but for this figure the waistline curve needs to be reshaped so that the deepest part will be at B. (See sketch.) This will eliminate the long look of the back.

It would be an expensive alteration to fit the rest of the jacket back, because the collar and sleeves would have to be removed to cut away the excess fabric, as the sketch illustrates.

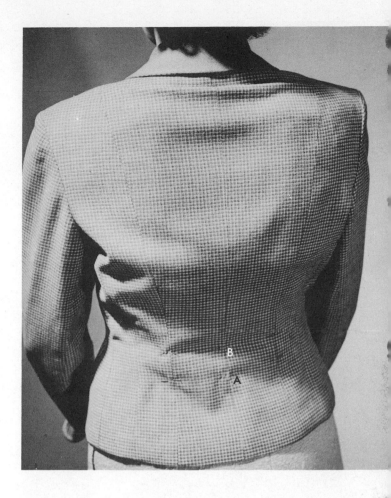

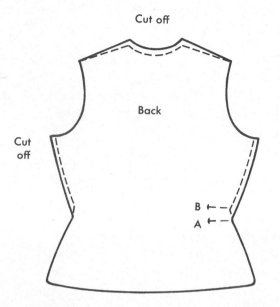

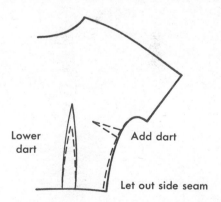

13. The sleeve length is attractive on this arm. The main fitting problem with the dress is that it is too tight in the bust. First, the bodice dart (A) should be lowered to end below the crown of the bust. It is robbing the figure of fabric that is there. Then, a bust dart must be added. The side seams of the bodice will still meet, because the back bodice will be shortened, as the next sketch points out.

Considering how tight this bodice is in the bust by the way it pulls at the buttons and buttonholes, it would not even be advisable to buy it unless there was enough seam to let out under the arm.

14. The way this dress wrinkles high and low in the back is typical of much of the look with ready-to-wear. To give more width across the back, the darts should be changed to two tucks. The extra length in the bodice would be cut off the bottom. The side seams of the bodice will still meet, because the front bodice needs a bust dart added, as the former sketch illustrated.

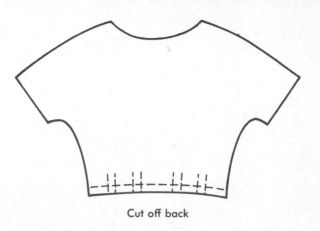

Cut off back

15. If a dress fits properly in the front of a bodice, but the back bodice is this much too long, do not buy it, because it could not be altered satisfactorily. In the preceding photograph and sketch, the alteration was feasible because the front of the dress needed a bust dart.

When shopping for ready-to-wear, it can be discouraging if the figure needs two sizes larger for the hips than the bodice. Sometimes, the side seams of the skirt can be let out enough, as the first sketch on the next page illustrates.

Do not buy a dress two sizes larger to fit the hips if the bodice must be made that much smaller. Such an alteration would be expensive and is not even recommended.

16. The front of this skirt is too tight in the hips, just as the former photograph illustrated with the back. The side seams can be let out, if there is sufficient allowance, and the extra width at the waistline put into more gathers, or extra tucks or darts if the dress had them instead of gathers.

The bustline is too low on this figure for the curve of this dress, so the curve needs to be relocated.

At A, the front of the sleeve needs to be dropped and cut off to straighten the line of the sleeve. Then, the sleeve would be hemmed again.

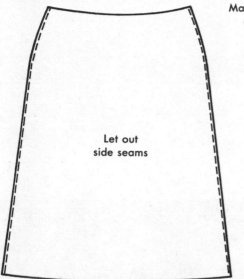

Make extra gathers, darts, or tucks or deepen present ones

Let out side seams

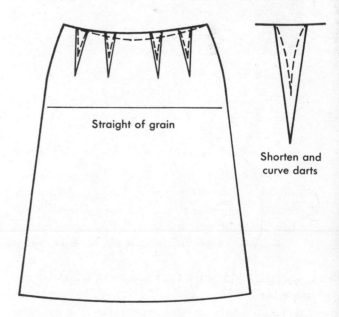

Straight of grain

Shorten and curve darts

17. The way this skirt cups at the back is a typical American look. To straighten the grain at the hipline, as the sketch illustrates, the skirt needs to be lifted at the center back. The right hip is even higher than the left hip. Then, the darts should be shortened and curved in stitching.

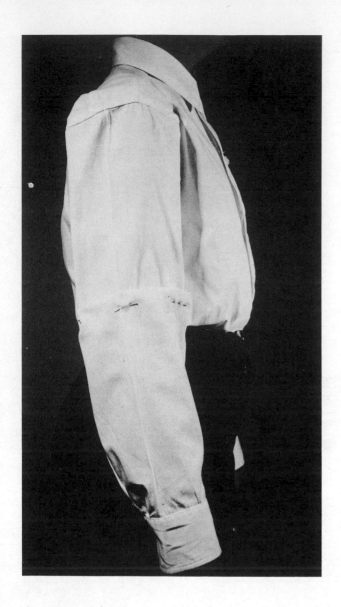

18. The correct sleeve length for a shirt is judged when the arm is down; there should not be a break in either the cuff or the sleeve. The tuck that is pinned in this shirt is to show how much the sleeve needs to be shortened.

Never take off the cuff to shorten the sleeve; always take out the sleeve at the armhole and cut off excess length at the top of the sleeve.

On less expensive shirts, it may be necessary to take in the underarm seam of the shirt (see sketch) so that the smaller cap will fit the armhole, or to add little pieces to the sleeve underarm. However, more expensive shirts have ease on the sleeve cap, and neither of these may be necessary, depending upon the amount the sleeve is shortened.

19. The back of this dress is entirely too short waisted. If there is sufficient seam allowance in the blouse, let it down as much as possible. The only other alternative is to wear a wide belt to cover the short-waisted look.

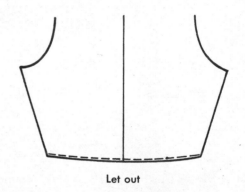

Let out

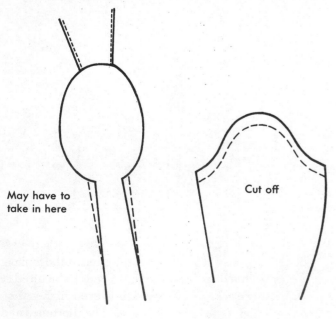

May have to take in here

Cut off

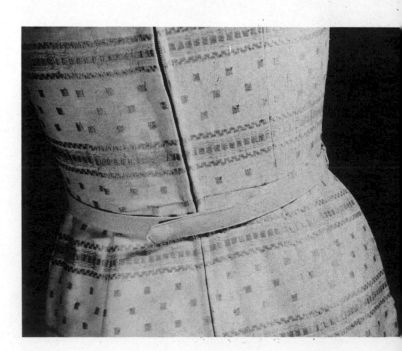

20. Two alterations are necessary on the front of this skirt. First, the front needs to be lifted and cut out at the center, tapering to nothing at the side seams. Then, the skirt will be more becoming on the figure if the fullness is arranged in two tucks or folds instead of gathers, and if they are placed closer to the front of the skirt.

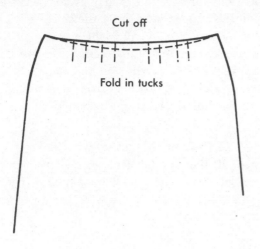

21. This figure definitely has a high right hip. If this skirt were being made, the pin on the waistband shows how much would be added to the skirt at the right top in cutting, tapering to nothing at the center back. However, in altering ready-to-wear, the only thing that can be done to straighten the skirt is to lift the left side as the sketch illustrates. The bottom of a pleated skirt must always be hemmed on grain, just as we say of a gathered skirt on page 112, *BMCC.*

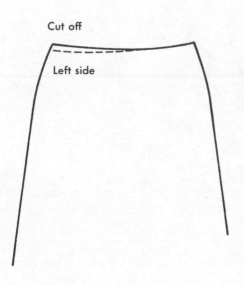

22. The Ivy League style of this dress is very popular, and this tight look at the waistline is common with it. Some of the fabric in the tucks should be let out, and some ease should be allowed on the bodice when it is restitched to the skirt. The wrinkles underarm will disappear when the bodice is loosened below. The remaining fabric from the tucks should be made into two smaller tucks on each side, and they should be closer to the side seams. If a figure has prominent shoulders, a bloused line at the waistline is always more becoming.

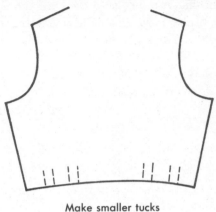

Make smaller tucks
toward side edges
Ease to skirt

23. This photograph illustrates another typical ready-to-wear look in the fit of the bodice. The bust dart (A) needs to be relocated so that it is in line with the crown of the bust. Then, the bodice darts (B) must be lowered so that they end under the crown of the bust. The midriff is too tight; so the bodice darts should be decreased at the same time they are being lowered, and the side seams of the front bodice let out. Then, the bodice can be eased to the skirt when the waistline seam is restitched.

There is enough width in the skirt below the darts, but the darts should be decreased so that the skirt would be eased when restitched to the tape at the waistline. This will eliminate the tight look in the skirt front below the belt.

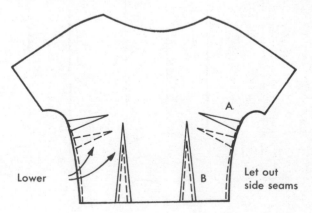

Lower

Let out
side seams

24. Trousers in the correct length do not have any break at the top of the cuffs. To alter the length of them, remove stitching and crease marks in present cuffs. Turn the bottom to the right side (A) at the correct length. Pin evenly all around and then press this line.

Measure up 1½ inches from this line and fold down (B). This line is the edge of the cuff, and should be pressed, also. Turn under the remaining length to the inside (C) and stitch to the underside of trousers. Press again (D) and the cuff is finished.

Turn down the cuff at the sides (E) and hand-stitch the cuff to the trousers about ½ inch from the top of the cuff.

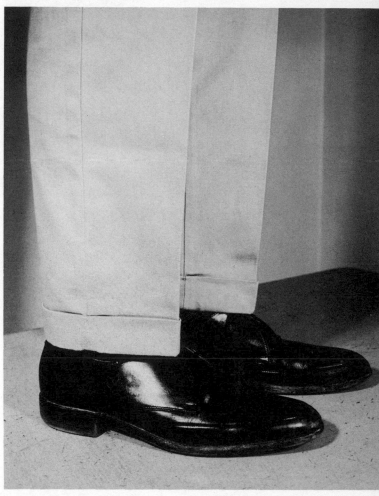

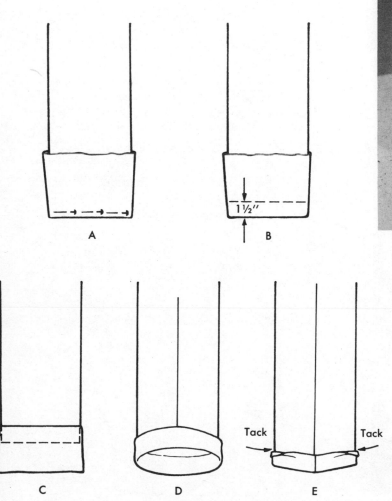

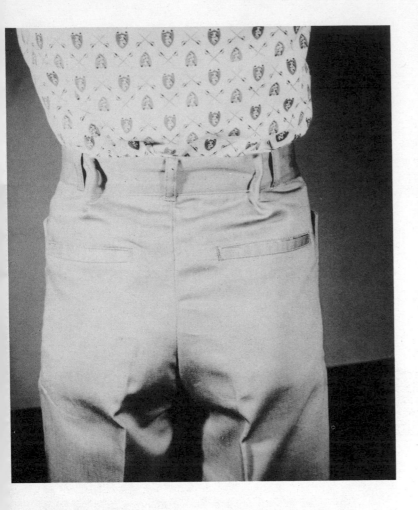

25. When trousers are too large at the waistline, they may be altered in two ways. First, there are normally darts in the design, which can be deepened where the tucks are pinned. If not, darts can be stitched on the inside to take out the excess at the area where the tucks are pinned. This is done on children's trousers so that the darts can be let out easily as the child develops.

The other method, and the one usually followed for an adult, is to make the alteration at the center back seam, unless it is so much that it pulls the trousers out of line.

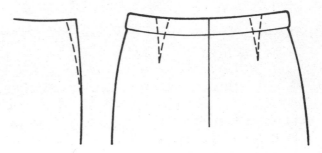

26. Any style of jacket should always hang with perfect straight of grain at the two front edges, as the sketch indicates. This jacket is being pulled out of line because the hips are large. The darts should be decreased—maybe even let out altogether to the waistline and just be stitched above. Then, if necessary, the side seams may also be let out.

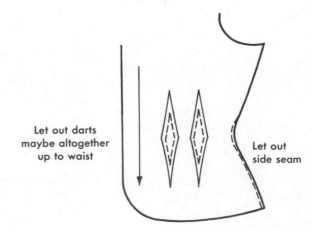

Let out darts maybe altogether up to waist

Let out side seam

27. Even if there were sufficient fabric in the seam of this sleeve to let it out enough for a comfortable size in the upper arm, it would be an expensive alteration in a lined garment.

To shorten the sleeve the amount indicated by the pinned tuck, the cuff would have to be removed. This would also be an expensive alteration.

Neither of these is recommended.

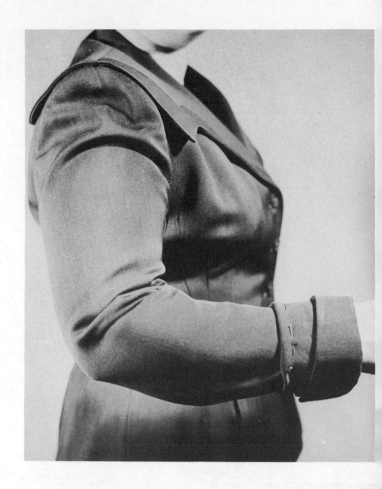

28. The darts are entirely too high and too long for this figure, and they are too deep. Notice the fullness at both ends of the darts. When they are decreased in stitching, the back side seams can then be made deeper to take in the excess fabric. The line of this dress would have been improved with a center back seam.

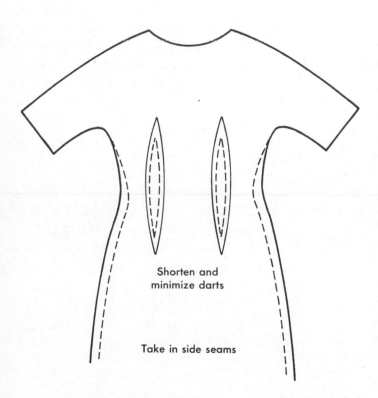

Shorten and minimize darts

Take in side seams

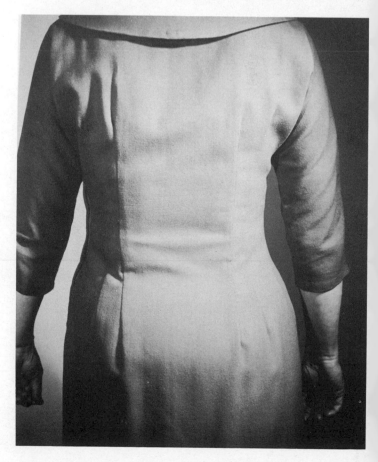

29. The way this shoulder pulls up at the front is a common shoulder problem. The front shoulder seam needs to be let out, but it would not be a recommended alteration for this dress because the collar line would never be right at the front. Always study the line of your garment to analyze if it can be altered. Never buy a garment that has straight of grain (A in sketch) at the shoulder line of a cut-on sleeve, as this dress does. It will never hang right when your arms are down. The shoulder line of the sleeve should always be cut with a bias line (B in sketch).

When ready-made clothes do not fit, it is due to our figure differences from the model figure, or to the poor cut of the garment. This dress has examples of both of these.

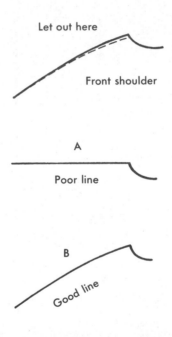

30. The detail in the design of this dress, with the fan-shaped darts at the bustline, would make it impossible to alter this dress through the bustline and have a quality look. It is not recommended.

Complete or back drums in skirts or dresses

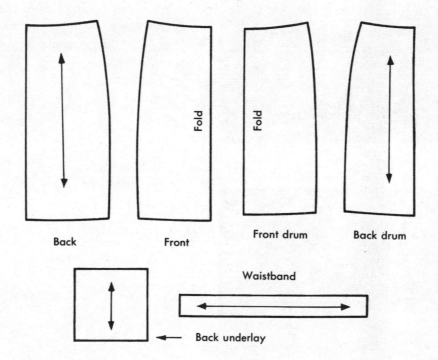

A seam opening in the back of a skirt with an underlay on the skirt drum has replaced pleats for walking room. The seam opening is smarter looking and is easier to maintain. It also reduces bulk in many fabrics, and it does not pull out of shape as pleats sometimes do.

With a fabric like wool, always use a complete drum. With fabric like cotton, just the back drum is preferred because of ease in laundering.

The complete or back drum is recommended in straight skirts, not only for the underlay at the seam opening, but for several other reasons. It helps skirts to retain their shape, gives more character to the appearance of the finished garment, and many times eliminates wearing a half-slip. A blouse or camisole is all that is needed with a suit skirt that has a complete drum.

Complete drums are made from firm rayon fabrics; a twill is a good choice. Suracel is highly recommended. Back drums are made from super soft or soft siri, or from one of the many synthetic, cotton, or silk underlinings on the market.

Drums are cut from the same pattern pieces as the skirt and on the identical grain. It is worthy of mention that a selvage could never be used at the lower edge of a drum, as you may sometimes find in ready-to-wear, because the grain would be entirely incorrect.

The complete drum

Skirt front unit. Staystitch waistline edge and sides of skirt from notch up just outside the seamline and in the direction arrows indicate.

Skirt back unit. Staystitch waistline edge and right side of skirt from notch up just outside seamline and in the direction arrows indicate. Staystitch left side of skirt from notch up ¼ inch from edge.

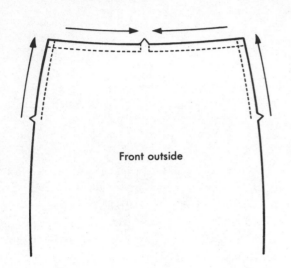

Front outside

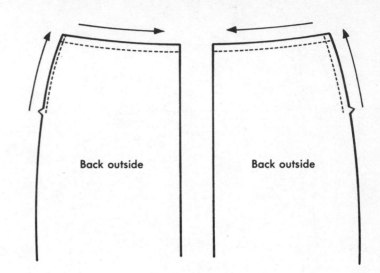

Back outside   Back outside

With right sides together, baste-stitch along center back line from lower edge for the desired length of seam opening. Place a square of seam binding over end of seam opening, and changing to regulation stitch, lockstitch at end of opening through seam binding. Continue stitching along center back line to waistline. If A edges are not selvage, staystitch ⅜ inch from edges and pink with pinking shears. Stitch darts and press toward center back.

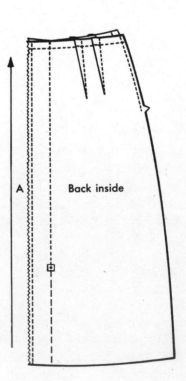

A   Back inside

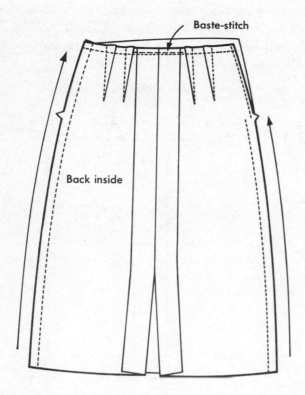

Baste-stitch

Back inside

Complete skirt unit. Press back edges open and baste-stitch to waist edge outside seamline. Remove baste-stitching below lockstitching at lower edge. Stitch side seams from lower edge to waist edge, leaving left side open exact length of metal part of zipper with tab turned up, plus waist seam allowance. Finish seams, if necessary, and press open.

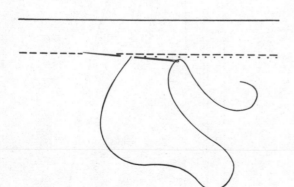

Insert zipper with the same learnings as on pp. 40–41, *The Bishop Method of Clothing Construction*.° Use hand-picking on front edge of placket instead of machine stitching.

You may baste-stitch last row of stitching on front edge as a guide for hand-picking. Then, baste-stitching can be removed when hand-picking is completed.

Do hand-picking from topside of skirt. Use buttonhole twist thread or double mercerized thread in needle. Draw thread through a beeswax holder. Bring needle to topside of fabric. Take stitch backward when putting needle to underside of fabric. Bring needle to topside of fabric about ¼ inch from first stitching. Continue hand-picking placket opening for a fine custom look. Remove baste-stitching.

*Throughout the remainder of this chapter this text will be referred to as *BMCC*.

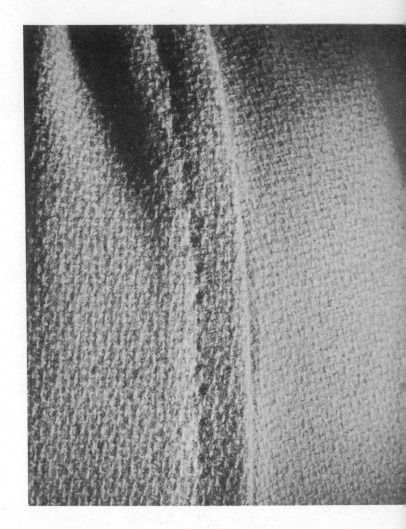

This photograph shows the completed placket with hand-picking after the baste-stitching has been removed.

Open out back extension edges and using hem gauge, turn up hem and press. Finish hem like photograph on page 171, *BMCC*. If the edges of the seam opening were not cut on the selvage, apply a piece of seam tape turning under lower ends. Slipstitch inner edge to hem.

If you are making a dress, the drum will have to be staystitched to the skirt at the waistline (see steps to follow), and the bodice joined to the skirt, before the zipper can be inserted. For approximately 1½ inches from the zipper opening, snip drum to seamline at waistline, and keep free from waistline seam until zipper is inserted. Then, the drum may be hemmed by hand around the placket (page 103, *BMCC*), or finished the same as the technique that follows here.

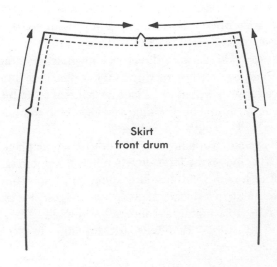

Skirt
front drum

Front drum unit. The drum is cut one inch shorter than the skirt. Staystitch waistline edge and sides of drum from notch up, just outside the seamline and in the direction arrows indicate.

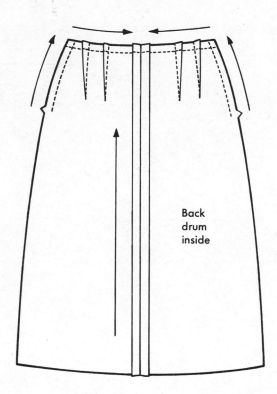

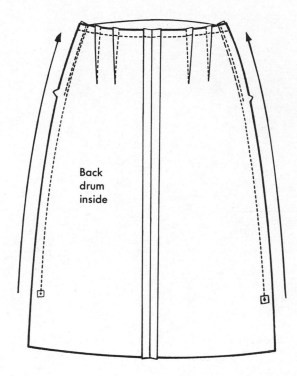

Complete drum unit. Staystitch waistline edge and sides of drum from notch up, just outside the seamline and in the direction arrows indicate.

With right sides together, stitch center back seam in direction shown. Finish with pinking shears, if necessary (this is cut on the selvage, when possible), and press open. Stitch darts and press toward side edges, opposite of the skirt darts.

Stitch side seams in direction shown, leaving them open at lower edge for three inches plus hem allowance for walking room. Place a square of seam binding at beginning of stitching, and lockstitch through it for added strength at opening.

Baste-stitch opening on right side for zipper placket; it will come on the left side when lining is put in skirt.

Finish seams with pinking shears, and press open. Stitch a strip of seam tape to raw edges of zipper opening; turn under raw ends at lower edge. Hem by hand to drum. Remove baste-stitching at placket opening.

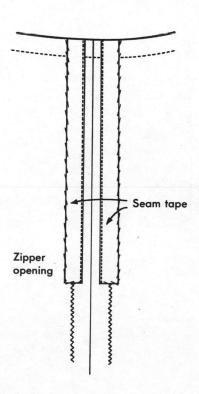

Seam tape

Zipper opening

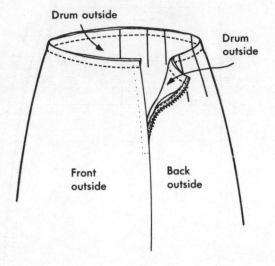

Drum outside

Drum outside

Front outside

Back outside

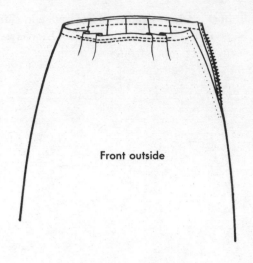

Front outside

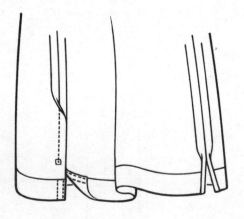

Complete skirt unit. With wrong sides together, pin drum to waist edge of skirt, keying centers and right side seam and placing side opening edges of drum over zipper on left side of skirt. Stitch waist edges together just outside seamline.

In the photograph on page 103, *BMCC*, the right side of the drum was placed to the wrong side of the skirt only because the dress was made from sheer wool and the drum was made from a firm taffeta.

Fold tucks on outside of skirt front through skirt and drum. Stitch tucks in place across waist edge.

Make the waistband and put it on the skirt exactly as you learned to do on page 170, *BMCC*.

Using hem gauge, turn up hem of drum and press. Finish like bottom photograph, page 46, *BMCC*. At side openings, stitch hem and drum together ⅝ inch from raw edge; staystitch raw edges ⅛ inch from edge. Press back seam allowance on ⅝-inch stitching line, turn under raw edges on ⅛-inch stitching line, and hem at machine, or slipstitch by hand. Drum will finish one inch shorter than skirt.

Complete or back drums

Skirt back underlay unit. Turn hem allowance on skirt back underlay to outside and stitch sides with ¼-inch seam for one inch from fold up.

Clip to end of stitching.

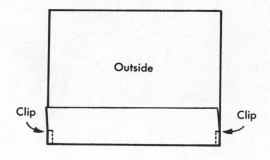

Turn hem to inside and press. Pink remaining edges. Finish hem same as skirt hem.

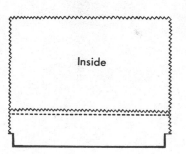

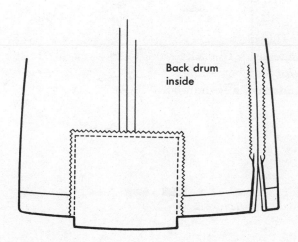

With wrong sides together, place underlay on back drum. Topstitch ¼ inch from edge. Underlay drops one inch below drum to finish same length as skirt.

When a complete drum is used in a lighter weight fabric, the underlay may be used all the way across the back drum, as it is when just a back drum is used (see page 47).

Final-press skirt (Chapter 9, *BMCC*).

This photograph shows the completed seam opening on outside of skirt. Underlay (A) is stitched to drum.

This photograph shows the completed drum and underlay. The skirt was turned wrong side out to take the photograph. When it is right side out, the underlay faces the wool skirt.

Back skirt drum.

The skirt back unit is prepared exactly like the directions on page 39, except that back darts are stitched later. Before sewing side seams, however, staystitch the back drum to the back of the skirt.

For approximately two inches above the seam opening at the back of the skirt, cut underlay of skirt fabric from same pattern as skirt back, and on identical grain. Omit the extension at center back and cut underlay on fold. Add ⅝-inch seam allowance to edge A, and cut off 1¼ inches from B edges. The underlay will finish the same length as the skirt. Cut drum area (C) from fabrics suggested at the beginning of this chapter. Cut like skirt back and on identical grain; add ⅝-inch seam allowance at A edge.

At B edges, staystitch ⅛ inch from edge. Turn in raw edges on staystitching and hem B edges at machine. One of the B edges may be cut on the selvage and will not require finishing. Turn up hem and finish same as skirt hem.

Staystitch A edges on underlay and drum just outside seamline. With wrong sides together, stitch ⅝-inch seam at A edge. Trim drum seam to scant ¼ inch. Staystitch seam allowance of underlay ¼ inch from raw edge, turn under seam allowance on staystitching, and stitch to drum in flat-felled seam.

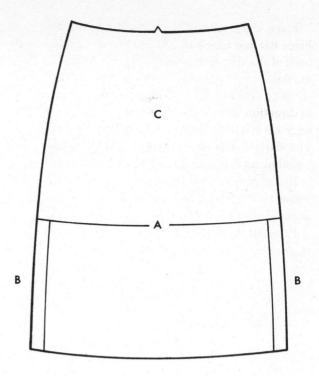

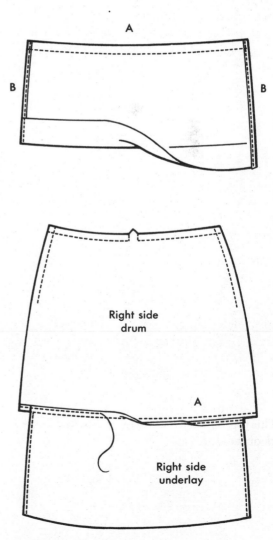

Place right side of drum to inside of skirt back. Place flat and check that it will fit and conform in the back of the skirt with perfection. Pin in place. Then, the darts can be stitched in the back of the skirt. Staystitch to skirt at D and E edges outside seamline, and in direction arrows indicate. Staystitch ¼ inch from edge on left side from notch up for zipper opening. The darts in the drum are just laid into place at the waistline as tucks, and are not stitched down.

If a skirt style has 3 gores in the back, then both seams could have a seam opening as we have shown at the center back.

Proceed with stitching side seams and remainder of skirt construction.

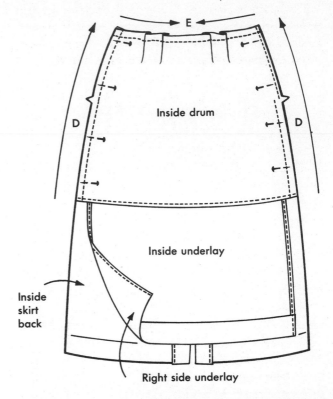

This photograph shows the completed drum in the back of the skirt.

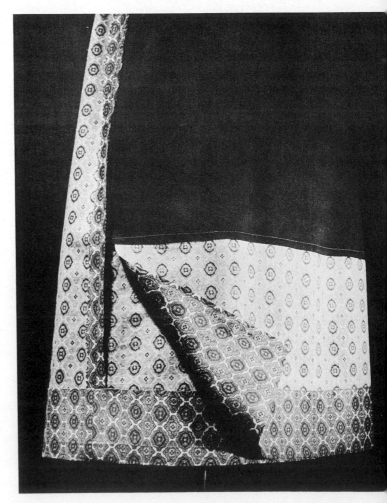

Lined toppettes for women and girls

Toppettes (or overblouses, as they are sometimes called) are popular, attractive, and comfortable, not only for children, but for women as well. They are worn over bathing suits, for a distinct change on a straight-lined or sheath dress, and with short shorts, bermuda shorts, pedal pushers, and slacks.

We have shown both styles (sleeveless and with cut-on sleeves) double and reversible. It is fun to make them reversible when the fabrics contrast on the two sides.

The double construction helps the overblouses to retain their shape, gives more support for adding various trims, and is more pleasing in appearance in lightweight fabrics. It helps us to attain our ever important objective—the quality look!

You may be able to cut several overblouses from just little pieces of fabric you have at home, and you can have fun making them in one evening!

A. Sleeveless toppette trimmed in wide rickrack. (This one is double, but not reversible because of trim.)

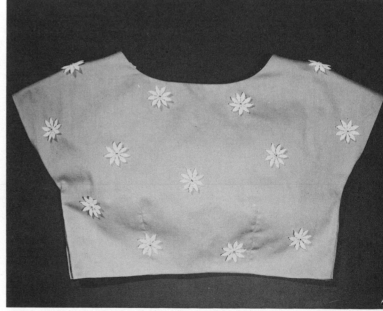

B. Toppette with cut-on sleeve. (This one is double, but not reversible because of trim.) A small pearl button is sewn in the center of each daisy.

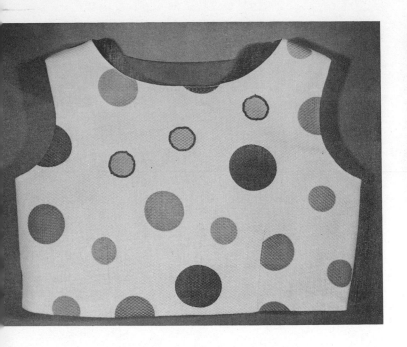

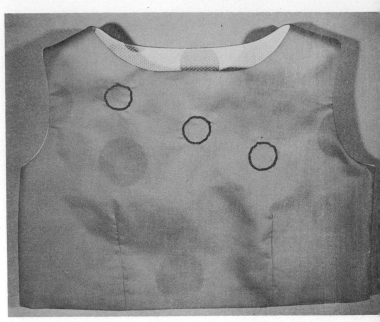

C. This is made of plain fabric on the underside, and is reversible. Circles are stitched with zigzag stitch.

D. This is the reverse side of C.

E. This is black fabric with a white print.

F. This is the reverse side of E, and is white fabric with a black print. These are known as companion prints.

Sleeveless toppette

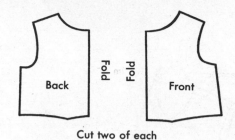

Cut two of each

Toppette A is cut 1¾″ longer at center front
C and D are cut like pattern

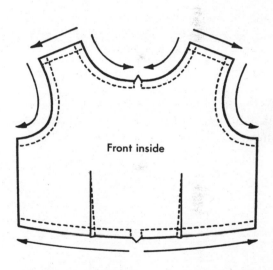

Front unit. Staystitch neckline, shoulders, arm-holes, and lower edge just outside seamline, and in the direction arrows indicate. Stitch darts. Repeat these steps for the second front.

Back unit. Staystitch neckline, shoulders, arm-holes, and lower edge just outside seamline, and in the direction arrows indicate. Stitch darts. Repeat these steps for the second back.

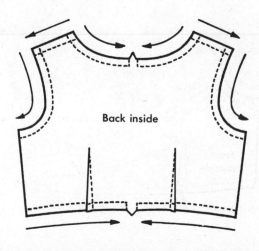

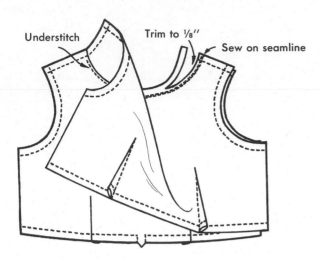

Complete toppette unit. With right sides together, sew one front to second front at neckline on seamline. Trim seam to ⅛ inch, and beginning and ending one inch from shoulder edge, understitch neckline.

As the photograph illustrates, to preserve the curve, work fabric up to needle in understitching this off-grain edge.

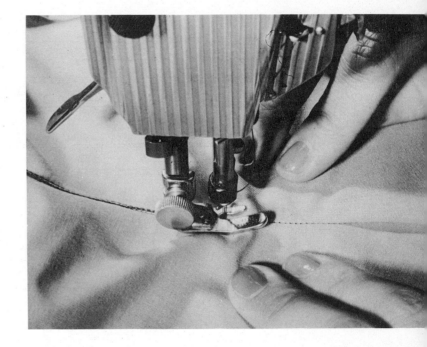

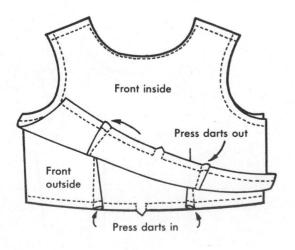

Turn fronts right side out, and press neckline. Press darts in one front toward center, and in second front toward side edge.

Repeat these same steps for the two backs of the toppette.

51

At armholes, right sides together, baste-stitch the second front to the first front for three inches down; lockstitch the threads; change length of machine stitch to a permanent one, and sew to side edge. Using an edge presser, press open baste-stitched areas of armholes. Trim the armhole seam to ⅛ inch. Remove baste-stitching and understitch lower part of armhole, ending one inch from side edge.

Repeat these same steps for the backs of the toppette.

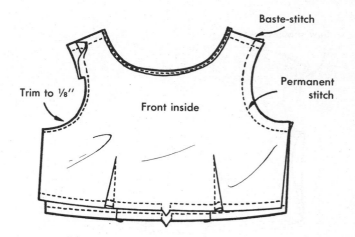

With right sides together, stitch first front to second front at lower edge on seamline. Trim seam to scant ¼ inch, and understitch, beginning and ending one inch from edge.

Repeat for backs of toppette.

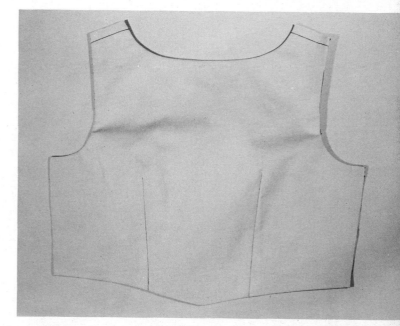

Turn backs wrong side out, and at the side seams, place right side of front toppette to right side of back toppette, and right side of front under section to right side of back under section. With a continuous line of stitching and a ⅝-inch seam, stitch a complete circle. Turn right side out after clipping seam to a point at armhole and lower edge. Press seam toward back of toppette (see this technique in photograph for vest, page 162).

Repeat for second side, except an opening must be left in the middle of underside to turn right side out. (See this step in photograph for vest, page 162.) Whipstitch these edges together.

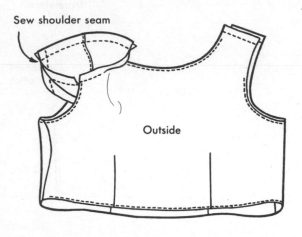

Sew shoulder seam

Outside

Closing the shoulder seams will now be a continuous operation with top and under section of toppette. On edge presser, press the seam open and trim to ¼ inch. Restore edges at neckline and armholes. Slip-stitch the pressed armhole edges together.

Final-press toppette (Chapter 9, *The Bishop Method of Clothing Construction* °).

This toppette was not made with double construction because the sailcloth fabric was too heavy for that technique. It has a facing cut all in one for neckline and armholes, as you learned to do on pp. 35–36, *BMCC*.

°This text will hereafter be referred to as *BMCC*.

Toppette with cut-on sleeves

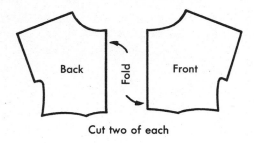

Cut two of each

Back unit. Make precisely like the front unit.

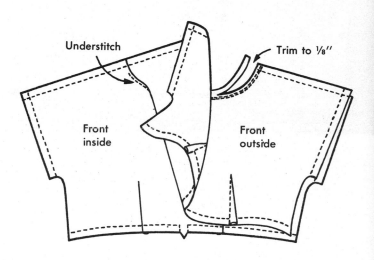

Complete toppette unit. With right sides together, sew one front to second front at neckline on seamline. Trim seam to ⅛ inch, and beginning and ending one inch from shoulder edge, understitch neckline. As the photograph on page 51 illustrates, to preserve the curve, work fabric up to needle in understitching this off grain edge.

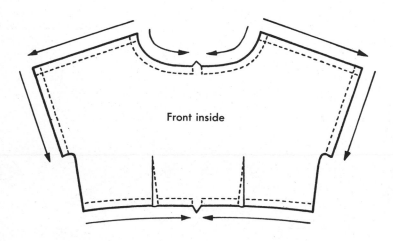

Front unit. Staystitch neckline, shoulders, sleeve edges, and lower edge just outside seamline, and in the direction arrows indicate. Stitch darts. Repeat these steps for the second front.

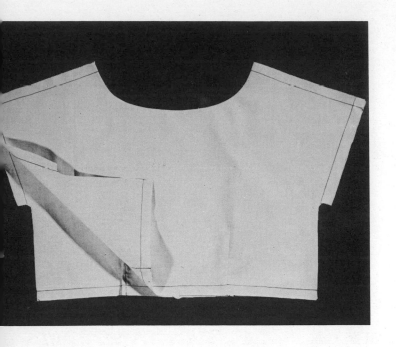

Turn fronts right side out, and press neckline. Press darts in one front toward center, and in second front toward underarm seam.

Repeat these same steps for the two backs of the overblouse.

Turn only the sleeve edges right sides together, and stitch sleeve edges on seamline. Trim seam to ⅛ inch, and understitch sleeve edges. Turn toppette right side out and press.

Closing the shoulder seams will now be a continuous operation with top and under section of overblouse. On edge presser, press the seam open, and trim to ¼ inch. Turn overblouse right side out and press shoulders.

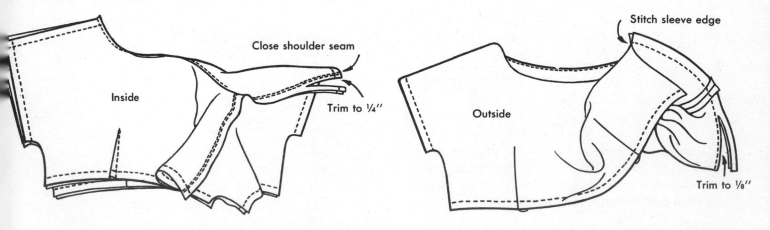

Inside

Close shoulder seam

Trim to ¼"

Outside

Stitch sleeve edge

Trim to ⅛"

Lined toppettes for women and girls

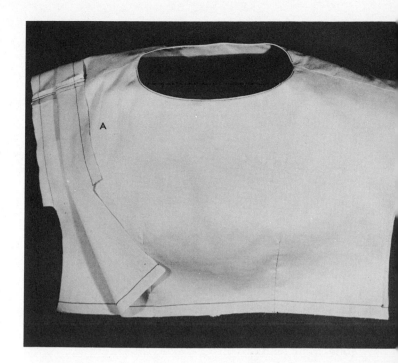

This photograph shows sleeve edge (A) before it was completed, and (B) after it was completed.

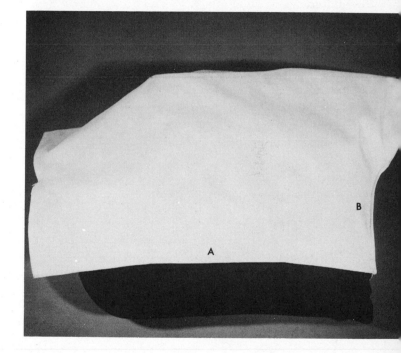

With right sides together, stitch lower edges of top and under sections of front toppette on seamline. Trim seam to ¼ inch and understitch (A).

Repeat these same steps for top and under sections of back toppette. Turn right side out and press lower edges of toppette.

At side edges, key together all four raw edges of fabric, and on the outside of toppette, stitch a ⅜-inch seam. Trim seam to a scant ⅛ inch and press together to one side. Turn to under sections and crease on stitched line. Stitch a scant ¼-inch seam from crease (B). Then, the seam may be stitched flat if the toppette is to be reversible.

Because of the cut-on sleeve, this style of toppette makes undesirable the technique that you used at the side seams of the sleeveless toppette (p. 52).

Final-press toppette (Chapter 9, *BMCC*).

Robes from quilted fabrics for women and girls

A robe from quilted cotton fabric will be cut and made in a different manner than a robe from any of the quilted synthetic fabrics because of the various characteristics of the fabrics. We are going to show the techniques on two robes for little girls, but they will be unchanged for all robes for young and old alike.

As we have said of children's dresses, not only do we want you to be able to turn out quality-looking robes, but we want them to be strong and durable for the hard wearings and many launderings they will receive.

Mother and daughter robes are always a family joy.

Before cutting out a robe, read the section of the chapter for the kind of quilted fabric you are using.

Basic learnings for a robe from quilted cotton fabric

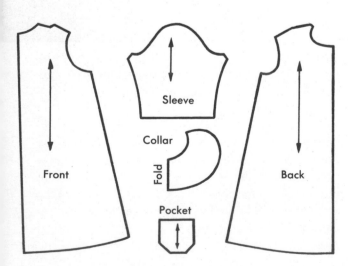

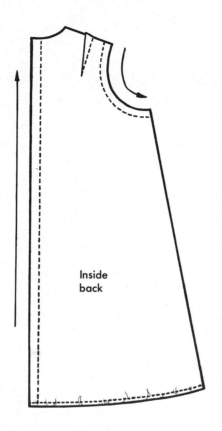

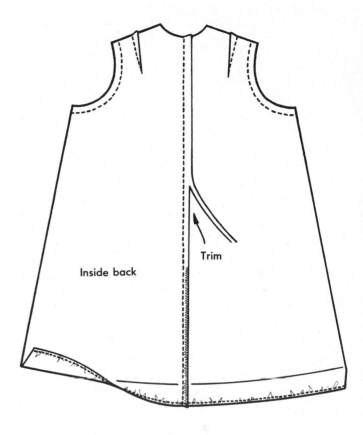

Back unit. Staystitch armholes just outside seam-line and in direction arrows indicate. The quilting helps to control the cut edges of the fabric, so it will not be necessary to staystitch any other edges.

Staystitch-plus lower edge (middle photo, page 46, *The Bishop Method of Clothing Construction**) ¼ inch from edge to prepare for hemming. With quilted fabric, use a short stitch at machine, and press finger as firmly as possible against the back of the presser foot to be certain of drawing up fabric as much as needed. If you ever draw it up too much, it is very easy to hold fabric firm and loosen after pressing hem. Stitch center back seam in direction shown, and stitch neckline darts.

Press center back seam to one side and neckline darts toward center back. Using hem gauge, turn up hem 1½ inches and press. Trim center back seam to ⅜ inch and zigzag edges together in same direction seam was sewn.

If you do not have a zigzag machine, a special attachment, called overcaster, is available for popular makes of straight-stitch machines. Or, you may choose to bind the seams as we will show on the front edges and around the collar. Do not trim any of the seam to the fold of the hem (see page 142, *BMCC*) in children's clothes because of the need to lengthen them.

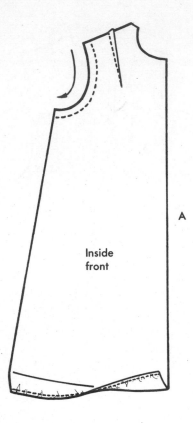

Inside
front

A

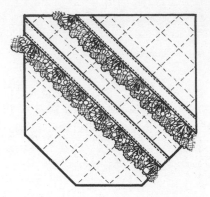

Pocket unit. In the Bishop method, we have repeatedly tried to instill the idea that *the fabric should tell you what to do* for a trimming detail on a garment. For this pocket, lace was gathered (see collar for instructions) and placed under edge of bias. Then, the bias was stitched flat at the machine, *following the design of the quilted fabric.* The top edge was bound (see collar for instructions) and the raw edges were pressed under ½ inch to form an attractively shaped pocket using the quilted design.

Front unit. A front facing is not recommended on the quilted cotton fabric. Trim away seam allowance at edge A (may be done to pattern before cutting out robe) to prepare edge for binding.

Staystitch armholes just outside seamline and in direction arrows indicate. Staystitch-plus lower edge same as back unit to prepare for hemming. Stitch neckline dart and press toward center front. Using hem gauge, turn up hem 1½ inches and press.

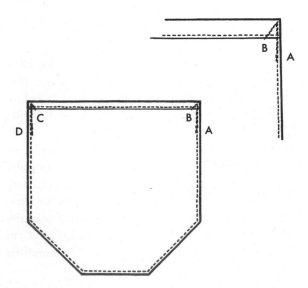

For added strength, make topstitching double at top of pocket. Begin topstitching at A, go up to B, turn robe and stitch all around pocket to C; turn robe again and complete stitching at D.

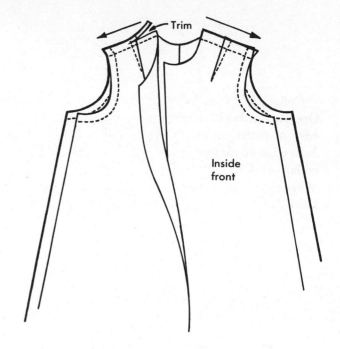

Complete robe unit. Stitch shoulder seams in direction shown, and press toward back. Trim to ⅜ inch, and zigzag in same direction as seams were stitched.

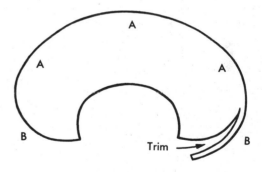

Collar unit. Trim away ⅝-inch seam allowance from outside edge (A) of collar to prepare for binding. If collar pattern has points at outer edges (B), round off for ease in binding.

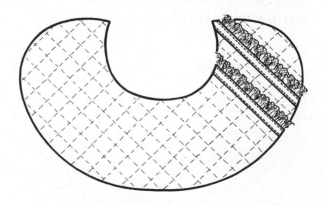

The trimming on the collar is a repeat performance of the pocket. Cut 1½ times the needed length of lace (for example, if you need 6 inches length, cut 9 inches), and gather by pulling a thread that is woven along the inside edge of lace. (See sketch on page 67.) Place lace under edge of bias, and stitch bias flat at machine.

Collar is always made singly in quilted fabric. Open up folded bias, and key raw edge of bias to raw edge of under collar. Stitch in fold of bias ¼ inch from edge on underside of collar. Turn bias to top of collar, and topstitch along second folded edge. Where and if necessary, trim away fraction of quilted fabric, so bias tape can lie smoothly for topstitching.

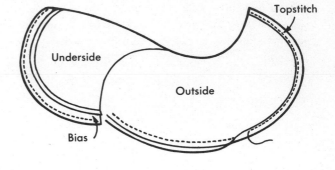

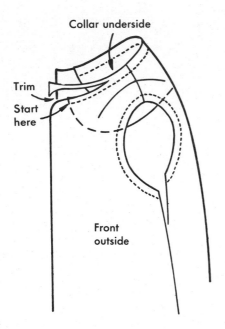

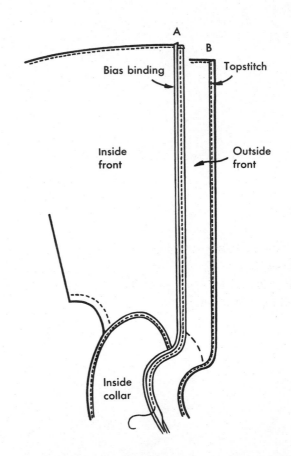

Place outside edge of collar to inside of robe. Key edges of collar to center fronts of robe, and center back of collar to center back of robe. With garment side up, stitch collar to robe in ⅝-inch seam. Trim collar and neckline seam to ¼ inch, rounding off corners to nothing at front edges.

At A and B, turn in raw edges of bias ¼ inch and beginning at one lower front edge of robe and continuing around neckline to second lower front edge, stitch bias tape to underside of robe, same as collar edge. Turn bias to top of robe and topstitch along second folded edge. Where and if necessary, trim away fraction of quilted fabric so that bias tape can lie smoothly for topstitching. Keep fabric to left of needle for topstitching.

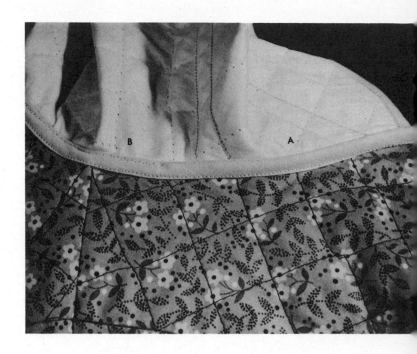

Turn folded edge of bias (A) to lie flat on under-side of collar. Stitch along folded edge of bias for added support (B) and smooth, flat finish on under collar.

This is the finished collar.

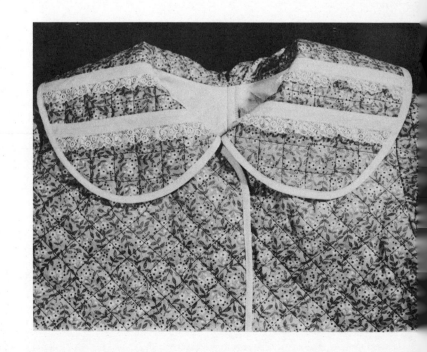

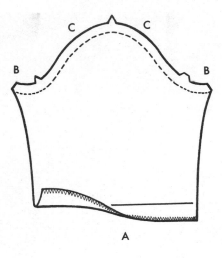

Sleeve and side seam unit.   Zigzag lower edge (A) of sleeve.   Precisely on ⅝-inch seamline, stay-stitch cap of sleeve (B and B), changing to longest stitch (C) between notches.

Using hem gauge, turn up hem 1½ inches and press.

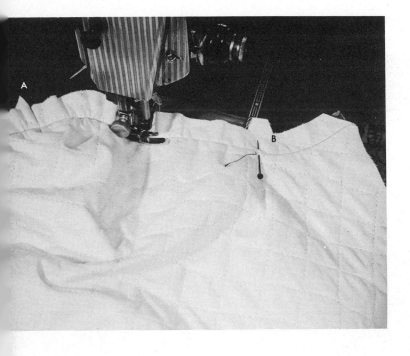

With right sides together, pin notch at top of sleeve (A) to shoulder seam of robe.   To prepare the sleeve to sew in the armhole, clip the first long stitch at the notch on the right side of the inside of the sleeve (B). Draw up that side of sleeve cap slightly to fit that half of the armhole.   Clip the first long stitch at the notch on the left side of the inside of the sleeve, and draw up second half of sleeve cap slightly to fit second half of armhole.   With sleeve side up, stitch sleeve to armhole of robe, sewing one thread inside line of stay-stitching and ease line.   Trim seam to ⅜ inch, and with sleeve side down, zigzag at machine.   Do not press (see photo, p. 70).

The sleeve and side seams of quilted robes are always stitched in one operation.   Stitch from bottom up (see photo, p. 70), and press toward back of robe. Trim to ⅜ inch, and zigzag together at machine from bottom up.

Hem sleeves at machine like bottom photograph, page 46, *BMCC*.

Hem unit. Press and restore hem lines at side seams of robe. Zigzag lower edge, and hem robe at machine. To hem by machine, fold the hem under as shown. Take 4 or 5 stitches along the underside of the hem. Then take 1 stitch into fold of the garment. Continue around entire hem.

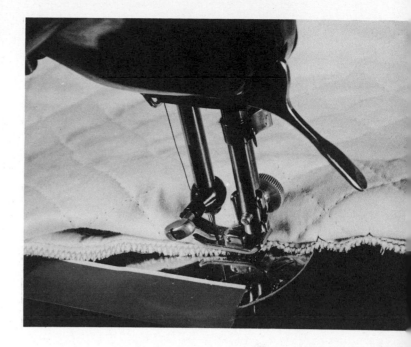

At front edge of robe, stitch hem firmly just inside bias tape.

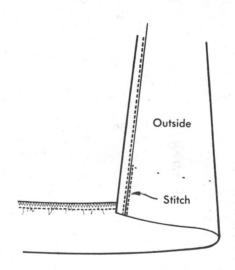

Complete robe unit. Make buttonholes and sew on buttons (page 84, *BMCC*). A small button on underside (A) is sewn with top button (B) for support to the single layer of fabric.

Final-press robe (Chapter 9, *BMCC*).

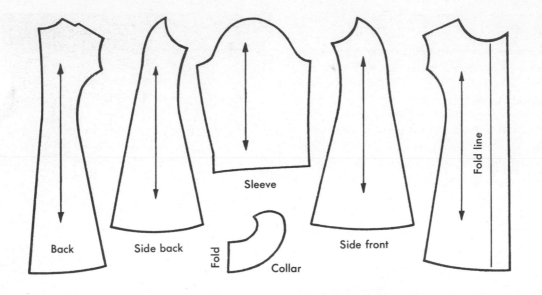

Back   Side back   Fold   Collar   Sleeve   Side front   Fold line

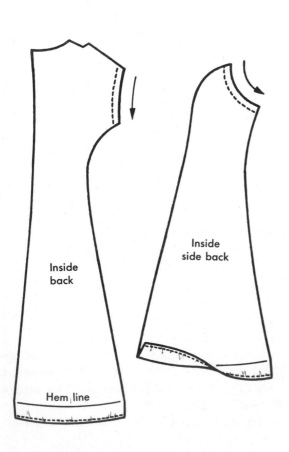

Inside back   Inside side back   Hem line

Basic learnings for a robe from quilted synthetic fabric

Back unit. Staystitch armholes of back and side back just outside seamline and in direction arrows indicate. The quilting helps to control the cut edges of the fabric; so it will not be necessary to staytitch any other edges.

Staystitch-plus lower edges (page 46, *BMCC*) ¼ inch from edge to prepare for hemming. With quilted fabric, use a short stitch at machine and press finger as firmly as possible against the back of the presser foot to be certain of drawing up fabric as much as needed. If you ever draw it up too much, it is very easy to hold fabric firm and loosen after pressing hem. Using hem gauge, turn up hem 1½ inches and press.

Stitch center back seam in direction shown, and stitch neckline darts. When attempting to stitch seams to come out thread perfect in this thick and slippery fabric, it will help to hold the under layer firm with your right hand. Stitch back of robe to side back, keeping *inside* curve on top. On right back seam, begin at bottom of robe, and keeping inside curve on top, stitch all the way to the armhole. On left back seam, stitch from bottom of robe to waistline for correct direction of grain. Then, turn and stitch from armhole to waistline with inside curve on top. On quilted fabric, you will have to clip the inside curve toward the seamline for stitching, because it does not give.

Press neckline darts to center back, press center back seam to one side, and press side back seams toward side edges. Trim seams to ⅜ inch and zigzag edges together in the same direction they were sewn. If you do not have a zigzag machine, a special attachment, called overcaster, is available for popular makes of straight-stitch machines. Or, you may choose to bind the seams with rayon or nylon bias tape as we have shown on page 61. Do not trim any of the seam to the fold of the hem (see page 142, *BMCC*) in children's clothes, because of the need to lengthen them.

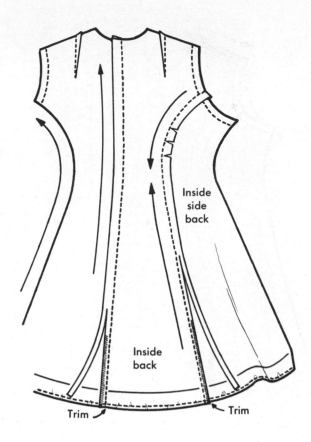

Front unit. A cut-on-facing is more desirable than a separate one with quilted synthetic fabric. If pattern has a separate facing, pin tissue together and cut as one.

To press, proceed as follows: Mark one layer of fabric only (A) with tracing paper and wheel. Through both layers of fabric, place 3 or 4 pins in marking line. Turn back and press top layer on pin line (see photographs, page 164). Remove pins. Turn fabric to have under layer on top. Fold back top layer to meet under fold and press second side.

Staystitch armholes just outside seamline and in direction arrows indicate. Staystitch-plus lower edge same as back unit to prepare for hemming. Using hem gauge, turn up hem 1½ inches and press.

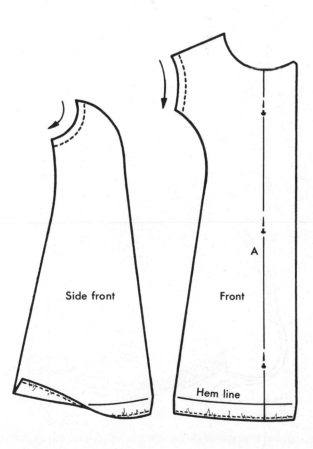

Stitch front of robe to side front, keeping *inside* curve on top, with the same techniques you learned on the back unit. On quilted fabric, you will have to clip the inside curve toward the seamline for stitching, because it does not give. Press seams toward side edges, trim to ⅜ inch, and zigzag edges together in the same direction they were sewn. Next, zigzag edges of facings (A).

Stitch shoulder seams in direction shown, and press toward back. Trim to ⅜ inch, and zigzag in same direction as seams were stitched.

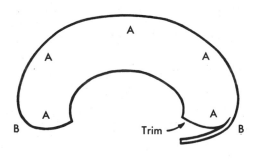

Collar unit. Trim away ½ inch of the ⅝-inch seam allowance from outside edge (A) of collar to prepare for binding (remaining ⅛ inch will be trimmed later). If collar pattern has points at outer edges (B), round off for ease in binding. Cut 1½ times the needed length of lace for A edge (for example, if A measures 20 inches, cut 30 inches of lace).

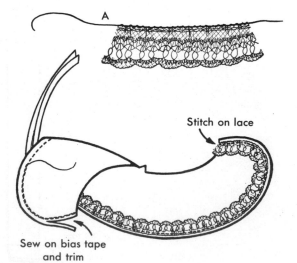

Gather lace by pulling a thread that is woven along the inside edge of it (A).

Stitch lace on collar ¼ inch from raw edge. Open out folded, commercial bias and right sides together, place raw edge along raw edge of collar. Turn to underside of collar and stitch on bias through row of previous stitching. Trim ¼-inch seam to ⅛ inch.

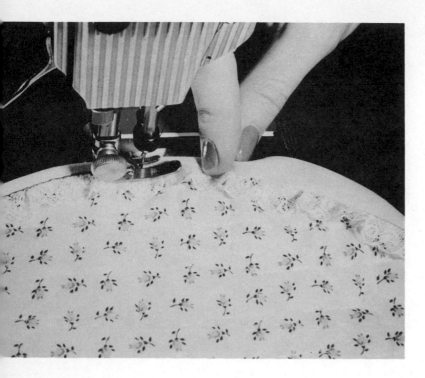

Turn bias to underside over cut edge, and without pressing it, stitch bias in place through seamline groove on topside. Collar is always made singly in quilted fabric. This lace trim is more unusual, attractive, and durable than having the lace turn toward the outside of the collar.

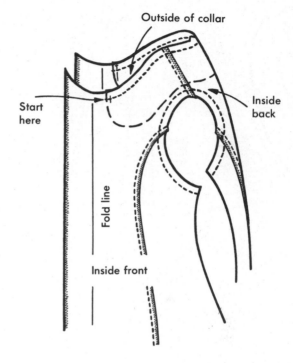

To put on collar, place underside of collar to outside of robe; key and pin outside edges of collar to center fronts of robe, and center back of collar to center back of robe. With garment side up, stitch collar to robe with 5/8-inch seam.

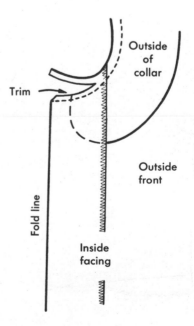

Then, turn facings to outside of robe on fold line, and again with garment side up, sew on facings through previous stitching to end of facings. Trim entire collar seam to 1/4 inch.

68

Open out folded, commercial bias, and with right sides together, place raw edge along raw edge of collar; overlap on facings for one inch (A). With garment side up, sew on bias through previous stitching.

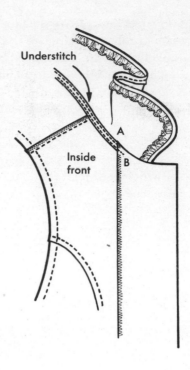

Understitch

Inside front

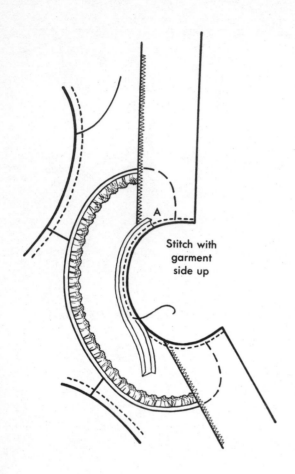

Stitch with garment side up

Turn facings to inside and understitch neck edge of bias (A). Then, stitch bias flat to robe, beginning and ending stitching ½ inch over edge of facings (B).

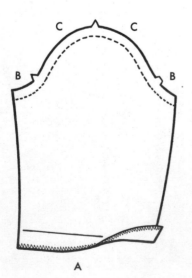

Sleeve and side-seam unit. Zigzag lower edge (A) of sleeve. Precisely on ⅝-inch seamline, staystitch cap of sleeve (B and B), changing to longest stitch (C) between notches.

Using hem gauge, turn up hem 1½ inches and press.

With right sides together, pin notch at top of sleeve to shoulder seam of robe. To prepare the sleeve to sew in the armhole, clip the first long stitch at the notch on the right side of the inside of the sleeve. Draw up that side of sleeve cap slightly to fit that half of the armhole. Clip the first long stitch at the notch on the left side of the inside of the sleeve, and draw up second half of sleeve cap slightly to fit second half of armhole. With sleeve side up, stitch sleeve to armhole of robe, sewing one thread inside line of stay-stitching and ease line.

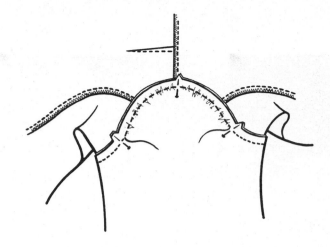

Trim seam to ⅜ inch, and with sleeve side down, zigzag at machine. Do not press.

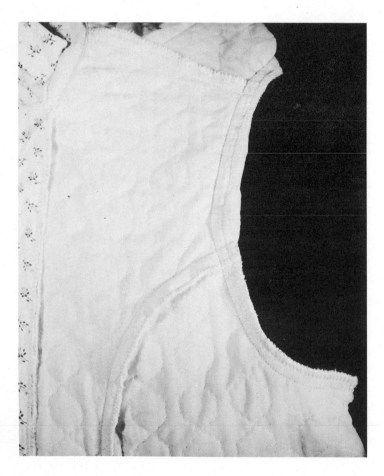

The sleeve and side seams of quilted robes are always stitched in one operation. Stitch from bottom up, and press seam toward front of robe, opposite of elbow ease, or possible elbow dart on pattern. Trim to ⅜ inch and zigzag together at machine from bottom up.

Hem sleeves at machine like bottom photograph, page 46, *BMCC*.

Hem unit.   Press and restore hem lines at seams of robes.   Zigzag lower edge.

Turn back facings and stitch hem of facings to hem of robe ¼ inch from edge.

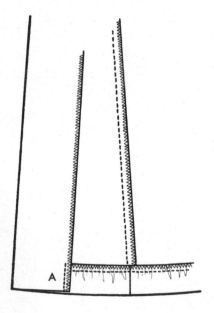

Turn facings to inside again, and stitch edge of facing to hem (A).

Hem robe at machine like bottom photograph, page 46, *BMCC*.

Complete robe unit.   Make buttonholes and sew on buttons (pages 84–85, *BMCC*).

Final-press robe (Chapter 9, *BMCC*).

Chapter **7**

Robes for men and boys

What a pleasant surprise it would be for your favorite men and boys to receive a robe you made for them personally. As you study the chapter, you will see that many fine techniques have been developed for making them. You will have fun in making robes, also. Popular fabrics are cotton broadcloth, cotton plaid, terry cloth, cotton or silk surah, seersucker, corduroy, or wool challis. The collar and cuffs may be in contrast, if desired.

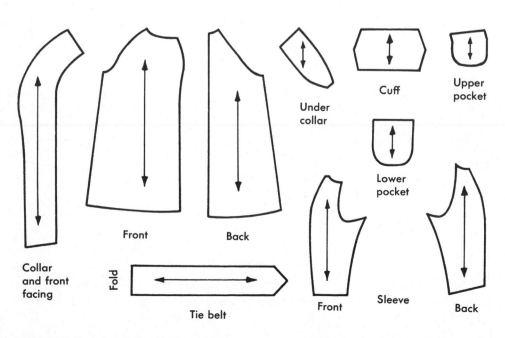

Collar and front facing

Fold

Tie belt

Front

Back

Under collar

Cuff

Upper pocket

Lower pocket

Front

Sleeve

Back

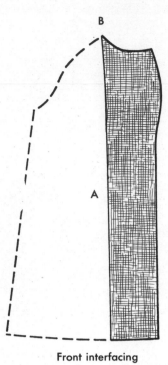

Front interfacing

Interfacing unit

Cut the interfacing from robe front pattern the width of facing pattern at A, tapering to edge of neckline at B.

Use the under collar pattern and cut the interfacing on identical grain of under collar.

With tracing paper and tracing wheel, mark a stitching pattern for stitching interfacing to under collar. It can form various designs, providing lines are on crosswise and lengthwise threads of interfacing.

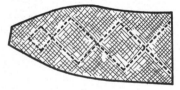

Collar interfacing

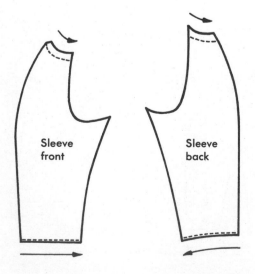

Sleeve front

Sleeve back

Sleeve unit

Staystitch neck edge of sleeve in direction shown, just outside seamline, and lower edge ¼ inch from edge.

Wrong sides together, sew sleeve back to sleeve front on seamline in direction shown.

Press seam toward sleeve front. Trim the lower seam allowance to ⅛ or ¼ inch; turn under raw edge of upper seam allowance, and stitch flat to garment in a flat-felled seam.

In plain fabric, the flat-felled seam is always finished from the topside of the garment, and the two rows of stitching give added interest.

However, if the fabric has surface interest, the flat-felled seam may be finished on the underside, and only one row of stitching will show on the topside.

With silk fabrics, use a French seam.

French and flat-felled seams are illustrated on page 120, *The Bishop Method of Clothing Construction.**

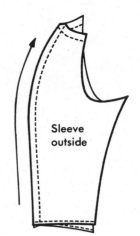

Sleeve outside

Back unit

Staystitch neckline in direction shown, just outside seamline, and lower edge ¼ inch from edge.

Turn under and press raw edge on staystitching ¼ inch from edge, and turn up hem desired amount. Press with hem gauge, shown on page 3, *BMCC.* Hem back unit like bottom photograph, page 45, *BMCC.*

However, in a cotton robe, the hem may just be topstitched to robe at machine.

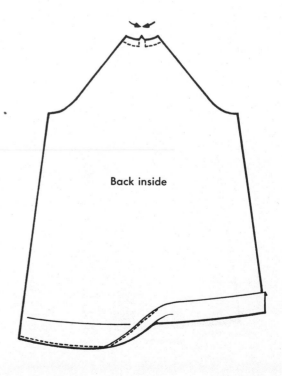

Back inside

*This text will be hereafter referred to as *BMCC.*

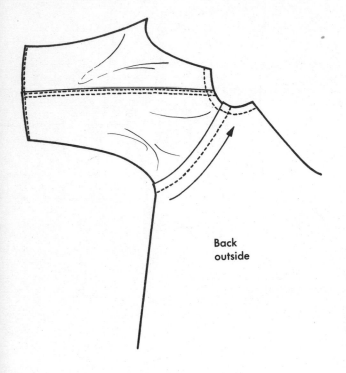

Back
outside

Wrong sides together, sew sleeve back to robe back
on seamline in direction shown.

Press seam toward sleeve and finish with a flat-
felled seam, the same as you learned to do in the
sleeve unit. Repeat for second sleeve.

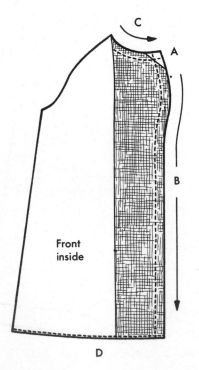

Front
inside

Front unit

Place interfacing on underside of robe front. At
corner A, trim interfacing out of seamline (page 173,
BMCC). Staystitch edge B on seamline in direction
shown, edge C just outside seamline, and lower edge
D ¼ inch from edge.

Turn under and press lower edge on staystitching
¼ inch from edge. Turn up hem desired amount and
press with hem gauge.

Pocket unit

The pockets are made and stitched to robe precisely like the instructions on page 121, *BMCC*.

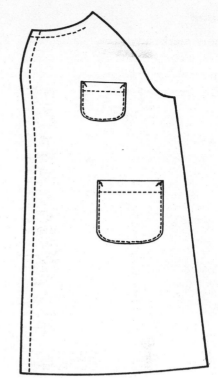

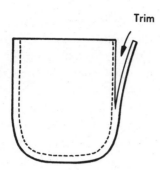

Trim

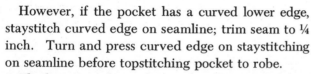

However, if the pocket has a curved lower edge, staystitch curved edge on seamline; trim seam to ¼ inch. Turn and press curved edge on staystitching on seamline before topstitching pocket to robe.

The letters on the pocket in the photograph at the beginning of this chapter were white, and were hand sewn to pocket with red embroidery floss.

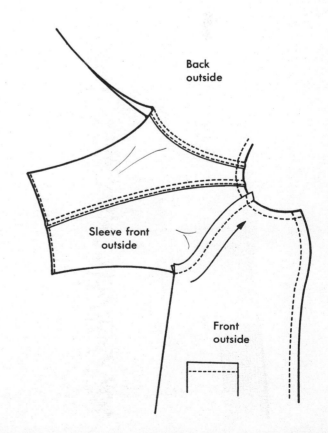

Back outside

Sleeve front outside

Front outside

Complete robe unit

Wrong sides together, sew sleeve front to robe front on seamline in direction shown.

Press seam toward sleeve and finish with a flat-felled seam, the same as you learned to do in the sleeve unit. Repeat for second sleeve.

Collar unit

At corners marked with A, cut away interfacing diagonally ¼ inch beyond seamline. Place interfacing with under collar, and staystitch B edges on seamline and C edges just outside seamline in direction shown.

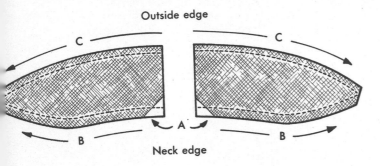

Outside edge

Neck edge

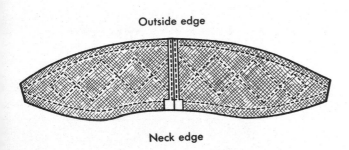

Outside edge

Neck edge

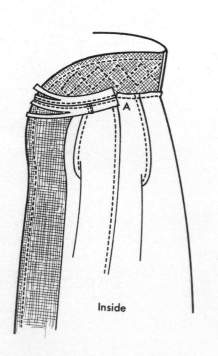

Inside

Stitch center back seam of under collar and interfacing and press open. Trim away interfacing up to seamline. Topstitch each side of seam, beginning at neck edge and continuing down one side to ¾ inch from outside edge. Cross over to other side and topstitch up to neck edge. Trim under collar seam to topstitching, except lower edge that is not topstitched.

The stitching pattern is drawn on *exact grain* of interfacing. Stitch on stitching pattern through interfacing and under collar. Do not sew beyond staystitching.

Right sides together, key center back of under collar to center back of robe. With robe side up, stitch collar to robe in a ⅝-inch seam.

Clip robe seam at two shoulder seams (marked with A), and press open seams up to clipping. Across the back of the robe, press the seam toward the collar. Trim interfacing out of entire seam to seamline, and trim seam across back of robe to ¼ inch.

Topstitch both sides of seam up to clipping at shoulder seam; trim seam to topstitching.

Facing unit

Right sides together, sew facing at center back on seamline. Press open seam and trim to ¼ inch.

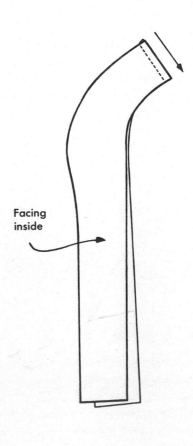

Facing inside

Right sides together, and with robe side up, sew facings to robe through staystitching on seamline of robe and under collar.   At outer edge, trim center back seam of under collar to ⅛ inch.

Trim interfacing to seamline and robe seam to ¼ inch.   In heavier fabrics, stagger width of robe seam as you have learned on page 161, *BMCC*. It is not necessary to reverse widths in staggering at the turn of the lapel in a robe.

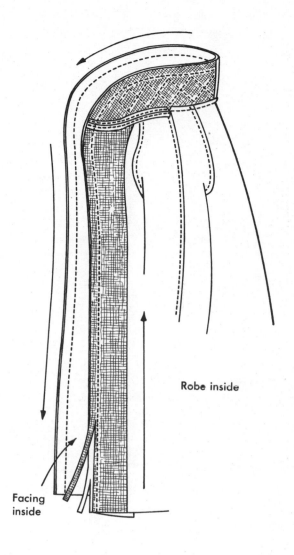

Robe inside

Facing inside

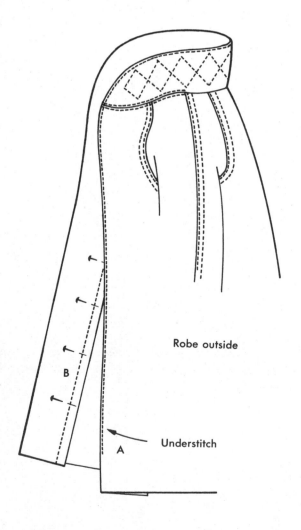

Robe outside

B

A   Understitch

Beginning and ending above the top of the hemline (A), understitch robe.   The understitching will be on the underside of the collar and lapels, but will be on the top of the robe for remainder of robe front below lapels.

Press this edge with perfection; then, pin edge of facing to edge of interfacing (B).   Staystitch together ⅝ inch from edge, and trim away interfacing to staystitching.

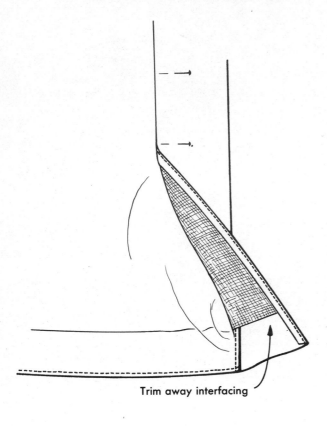

Trim away interfacing

Press under seam allowance on facing up to stay-stitching and pin flat to robe.

Restore hemline at lower edge of robe. Trim away interfacing a little beyond depth of hem.

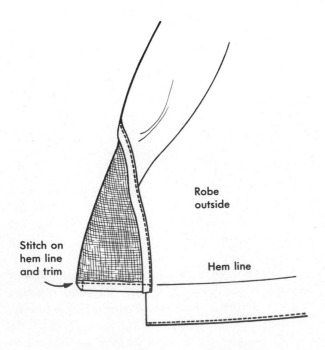

Robe
outside

Stitch on
hem line
and trim

Hem line

Turn facing to outside of robe and stitch on hemline. Trim away hem allowance to ¼ inch; turn right side out and press.

Finish hem same as back unit. Then, stitch the facing to the robe at the machine, going all around the facing continuously from lower edge A to the same place on the opposite side.

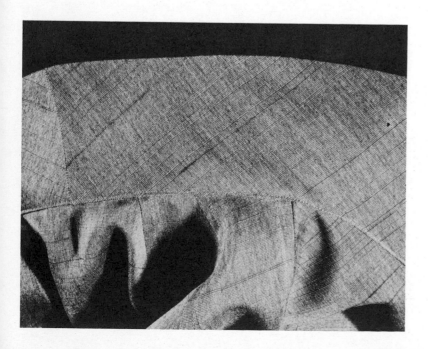

This photograph shows the facing as it forms the top collar. It is stitched to the inside of the robe at the machine.

Complete robe unit

Make a strip of bias for belt loops as you learned to do on page 193, *BMCC*.

Stitch in place at waistline, keying raw edges of loops to raw edges of back side-seam edge on both sides of robe back.

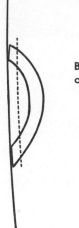

Back outside

Wrong sides together, sew side seams of robe and sleeves on seamline from bottom up. Press seam toward back of robe and finish with a flat-felled seam the same as you learned to do in the sleeve unit.

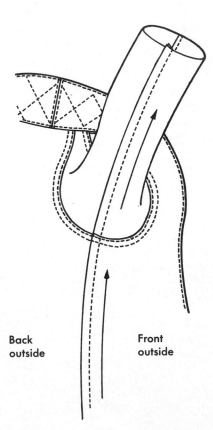

Back outside

Front outside

Right sides together, stitch a ¼-inch seam on belt, leaving an opening in the middle of the belt at the seamline to turn right side out. Trim corners, turn, and press. Slipstitch opening by hand.

Fasten belt to robe with a second piece of bias (A) that is placed through belt loop (B). Anchor-stitch piece of bias to belt at C and D. The belt has some give, but being attached it will never fall off the robe.

Cuff unit

Right sides together, sew cuff seam on seamline. Press open seam, trim to ¼ inch, and turn right side out.

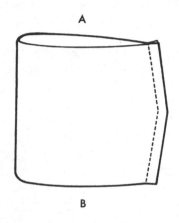

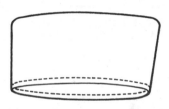

Staystitch A and B edges together (shown in previous sketch) ¼ inch from edge.

Place right side of cuff to wrong side of sleeve. Sew cuff to sleeve on seamline, and finish with a flat-felled seam exactly as you learned to do on page 107, *BMCC*.

Final-press robe, Chapter 9, *BMCC*.

Chapter 8

Children's dresses

A good deal of time and research has gone into the preparation of this chapter. Not only did we want you to be able to turn out quality-looking dresses, but we wanted the dresses to be strong and durable for the hard wearings and many launderings they will receive. The constant mending with a stitch here and there at the end of every laundering has been eliminated. We have used techniques, such as the finishes on the lower edges of the sleeves of the two dresses, that will make them easy to iron, also.

We will make two dresses in this chapter to cover all the techniques we want to present, but certainly you could combine and use some of the techniques from both dresses that would be desirable for the style and fabric you were using. It is worthy of note that there is still a continuing trend of mother and daughter fashions.

Remember that your daughter depends on you to teach her how to dress. She should be dressed with care, with sufficient attention paid to details, but without overemphasis on looks and clothes.

A child's clothing should reflect childhood—with gay colors, amusing touches now and then, but with simplicity. Colors should be becoming to hair, eyes, and skin coloring, and should coordinate well. Linens, broadcloths, fine cottons, and other simple fabrics are preferred to the overdone varieties.

To determine the pattern size for children's dresses, take a chest measurement, the same as the high bust measurement is taken for women. If necessary, pattern alterations are also made before cutting out dress. For example, the bodice would be lengthened or shortened the same as it would be done for women.

In purchasing fabric for children's dresses, it is most important to buy prints that are printed on grain, because the off-grain design would show when the strips are torn for the gathered skirt. Plain, dark cottons do not hold up as well in laundering as a small dot, check, or print. Then, unless it is removable, a white trim on a dark dress (such as a white collar) is not desirable.

Read the entire chapter before cutting dress.

Children's coats are made with the same techniques shown for women's coats in Chapter 11, *The Bishop Method of Clothing Construction.*

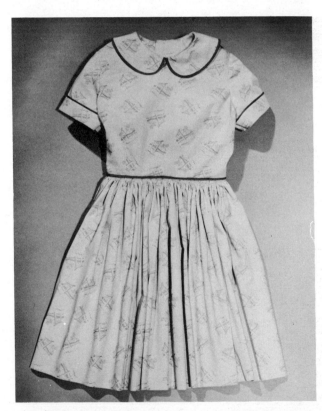

Basic learnings for a dress without underlining bodice

*Referred to hereafter as *BMCC*.

Children's dresses

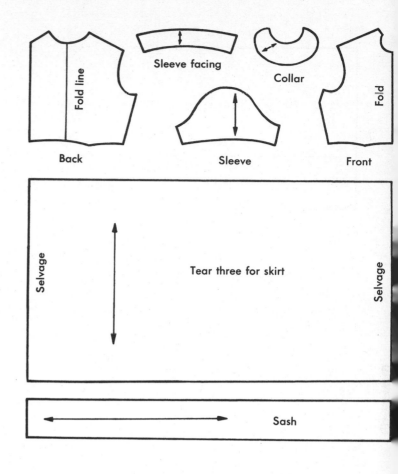

Back   Sleeve facing   Collar   Front

Fold line   Sleeve   Fold

Selvage   Tear three for skirt   Selvage

Sash

Skirt unit. Determine length to tear strips by the height of the child and desired width of hem. This first dress is a size 8 in a softer fabric, and the hem is 8½ inches deep. The dress to follow is a size 6 and the hem is 3½ inches deep. A deeper hem gives more character and better quality to the skirt.

Always tear the strips for a gathered skirt and always tear them to hang lengthwise. If desired, the top of the dress can be cut crosswise in a stripe, etc., but never the skirt for a quality look. If you buy a border print, the skirt should be pleated instead of gathered, because fabric does not gather gracefully on the lengthwise edge. Both dresses have 3 widths of fabric in the skirt, although a smaller size would only be 2 widths. This dress was made from fabric 45 inches wide, and the next dress was made from fabric 36 inches wide.

In both books, we have tried to stress that one of your keys to success is the preparation of your fabric to grain perfection. Press the torn strips with the iron to restore the grain on the torn edges, and also to assure that the skirt pieces lie grain perfect when folded in half.

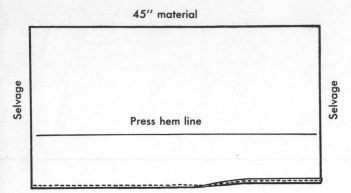

45″ material

Selvage

Selvage

Press hem line

One skirt section

Staystitch each skirt section ¼ inch from lower edge. Then, using hem gauge, turn up hem to correct finished length and press. Next, turn under raw edge ¼ inch on staystitching and press.

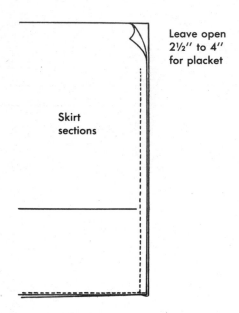

Leave open
2½″ to 4″
for placket

Skirt
sections

Open out hem and stitch seams of skirt ⅜ inch wide, leaving one of them open at top of skirt for 2½ to 4 inches for placket opening at center back.

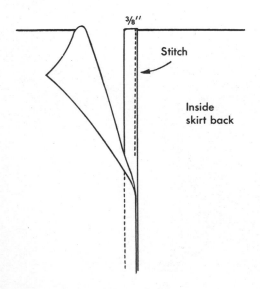

⅜″

Stitch

Inside
skirt back

On left side of center back opening, turn in ⅜-inch seam allowance, press, and stitch near edge of selvage.

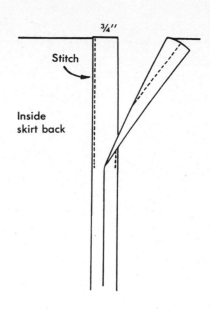

On right side of center back opening, turn in ¾ inch, press, and stitch near edge of selvage.

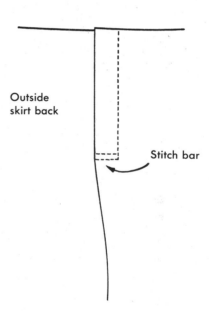

Lap right side of opening over left one, and stitch a bar at lower edge to hold opening securely. This technique is similar to the placket opening for shirts, page 108, *BMCC.*

Press open skirt seams and restore hem lines. Hem dress at machine as in bottom photograph, page 46, *BMCC.*

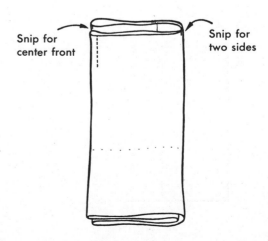

Beginning at center back, fold skirt in 4 sections. Snip with scissors to identify center front and 2 sides. This division will place more gathers in the back of the skirt than the front, as it should be, since the front bodice is cut wider than the back.

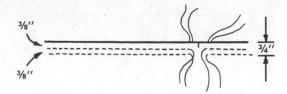

Outside skirt

Make gathers in 4 sections with longest stitch at machine. The first row will be ⅜ inch from raw edge, and second row ⅜ inch from first row (or ¾ inch from raw edge). You may find that using the attachment for gathering is very satisfactory if you can control the amount of gathers to fit the bodice.

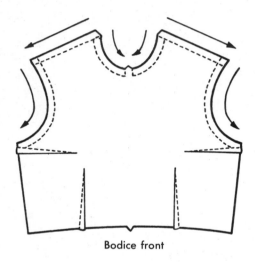

Bodice front

Bodice front unit. Staystitch neckline, shoulders, and armholes just outside seamline, and in direction arrows indicate. Stitch darts (page 27, *BMCC*), and press bust darts toward waistline, and waistline darts toward center front.

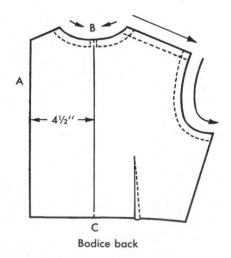

Bodice back

Bodice back unit. It is preferable to have facing cut in one with bodice back, and to have A edge on the selvage. Make facing wide for support; on this size 8 dress, we cut it 4½ inches wide.

Clip fold line for facing for ⅛ inch at B and C. Turn facings to inside at clips and press in place on fold line.

Staystitch neckline, shoulders, and armholes just outside seamline, and in direction arrows indicate. Stitch waistline darts and press toward center back.

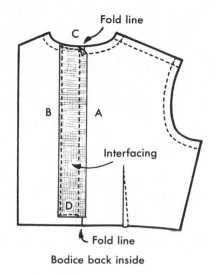

Fold line
C
B | A
Interfacing
D
Fold line
Bodice back inside

Tear a lengthwise strip of interfacing the length of bodice and 1½ to 2 inches wide to reinforce opening for buttons and buttonholes. Place it toward back of bodice (A) ⅛ inch beyond fold line, but stitch in place on facing side ⅛ inch from fold. Stitch ⅛ inch from second edge of facing (B) and neckline (C). Stitch ⅝ inch from lower edge (D), and trim away interfacing up to stitching.

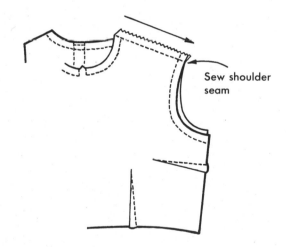

Sew shoulder seam

Stitch shoulder seams in direction arrows indicate, and pink with pinking shears.

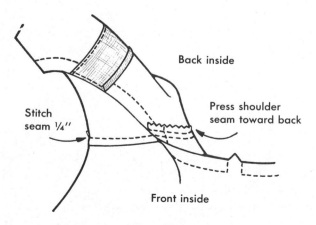

Back inside

Stitch seam ¼"

Press shoulder seam toward back

Front inside

Press toward back of bodice. Stitch seam ¼ inch from shoulder seamline to hold in place on back of bodice.

Collar unit. Make enough bias cording for collar, sleeves, and front of waistline seam. If the dress does not have a sash, bias cording is used around entire waistline seam. Cut true bias one inch wide (page 193, *BMCC*). Use a very fine cable cord, or a very heavy string, and fold bias over cord. Stitch close to cord with left side of zipper foot. For collar only, trim seam to perfect ¼ inch beyond stitching line. It is much easier to manage stitching cording on curved collar with a narrow seamline.

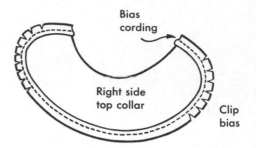

Bias cording

Right side top collar

Clip bias

Trim away ⅝ inch from outside edge of top and under collar. ⅜ inch is the difference in seam allowance given by the commercial pattern that you will not be using, and the ¼ inch is the amount collar will be larger than original pattern with bias cording extending on finished collar.

Place raw edges of bias to meet raw edges of *top* collar. Clip seamline of cording around curves of collar. Using left side of zipper foot, stitch through same line of previous stitching on cording.

With right sides together, and with short stitches at machine, sew under collar to top collar. Keep top collar side turned up, and sew through stitching on top collar, again with zipper foot.

Trim all of seam to scant ⅛ inch; turn, and press.

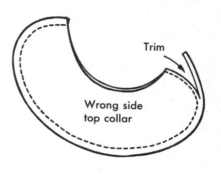

Trim

Wrong side top collar

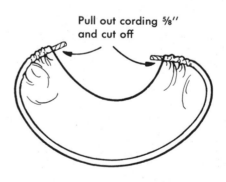

Pull out cording ⅝″ and cut off

At each end of collar, pull cording out ⅝ inch and cut off.

Then, restore shape of collar. At neckline, stitch together top and under collar just outside seamline. Key the two collars precisely and overstitch about 1 inch to hold together so that they will not slip or spread when attached to the center front of bodice.

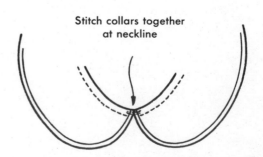

Stitch collars together at neckline

Place collar in position at center front and center back, and with dress side up, stitch collar to dress with ⅝-inch seam allowance. Clip curve of dress and collar frequently up to seamline to straighten curve for ease in stitching.

Staystitch shoulder edge of facing (A) ⅛ inch from edge, and turn in raw edge on staystitching to clean-finish.

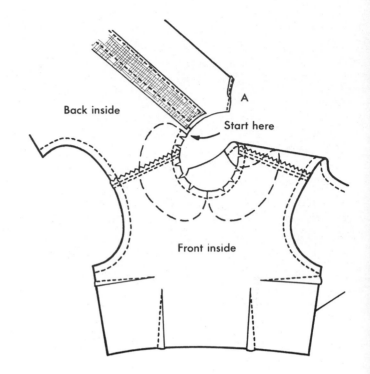

Turn facings to right side of dress on fold line, and stitch in place at neckline with dress side up. Trim entire neckline seam to *scant* ¼ inch.

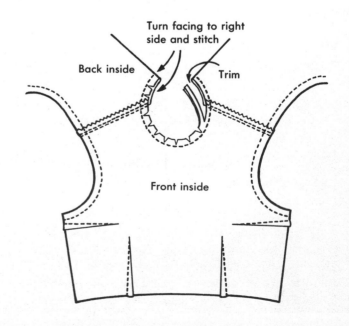

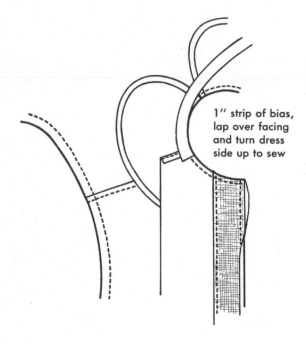

Cut a piece of bias one inch wide and lap over facings ¼ inch. With dress side up, sew on bias through previous stitching.

1" strip of bias, lap over facing and turn dress side up to sew

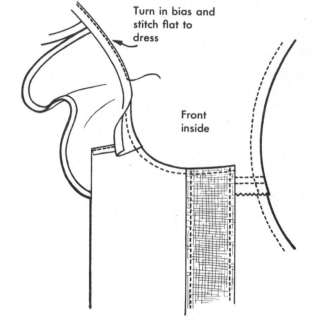

Turn in bias and stitch flat to dress

Front inside

If necessary, trim width of bias, so it will just turn in a narrow amount and fit over scant ¼-inch seam. Stitch flat to dress.

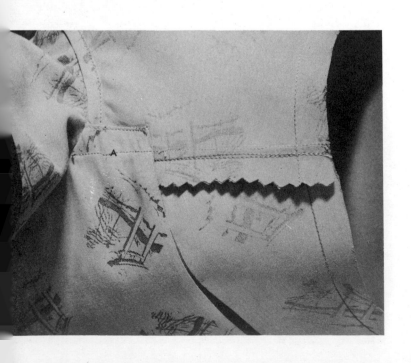

Turn facings to inside of dress. Then, for added strength, and with dress side up, stitch back facing in well of shoulder seamline (A). Press finished neckline. We have eliminated the front facing at the neckline for ease in ironing.

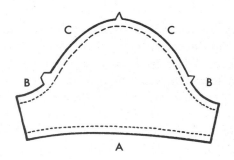

Sleeve unit. Staystitch lower edge A, just outside seamline. Precisely on ⅝-inch seamline, staystitch cap of sleeve (B and B), changing to longest stitch at machine (C) between notches.

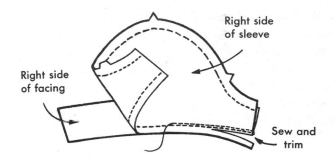

Place right side of facing to underside of sleeve, and with sleeve side up, sew facing to sleeve on seamline. Trim seam to scant ¼ inch. Understitch seam, beginning and ending one inch from edge. Turn facing to top of sleeve and press in place.

Trim seam of cording to perfect ¼ inch beyond stitching line. Trim away ⅜ of ⅝-inch seam allowance at top of facing. It is always easier to reduce seam when sewing in bias cording.

Place raw edges of bias to meet raw edge of facing. Using left side of zipper foot, stitch through same line of previous stitching on cording (A). Place facing in position on sleeve, press, and stitch facing to sleeve in well of cording seamline (B). At each end of sleeve, pull cording out ⅝ inch and cut off (C and C). Then, restore shape of sleeve.

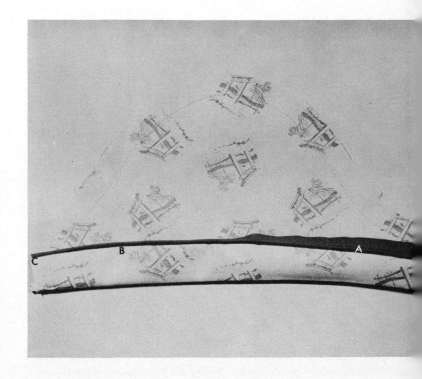

To prepare sleeve to set in armhole, clip the first long stitch at the notch on the right side of the inside of the sleeve. Draw up sleeve cap slightly to fit armhole, and fasten thread. Shrink out ease (p. 92, *BMCC*). With sleeve side up, stitch sleeve to armhole of dress. (See photos, page 63.) Sew one thread inside line of staystitching and ease line. Trim seam to ⅜ inch, and with sleeve side down, zigzag at machine. (See photo, page 70.) If you do not have a zigzag machine, make a second row of stitching ¼ inch from first row and pink near second stitching. Then, you may also wish to overcast by hand. However, a special attachment, called overcaster, is available for popular makes of straight-stitch machines.

With fabrics that are both sheer and crisp, such as organdy, the seams have a tendency to irritate a child's tender skin. To avoid this, spread the armhold seams and turn the edge of each seam toward the inside. Trim the seam allowance if it is too wide. Overcast the folded edges together. This makes a hand-turned French seam.

Complete bodice unit. Trim seam of cording to perfect ⅜ inch beyond stitching line to put in front waistline seam; this is the same width as first row of gathers from edge of skirt. Cording is used around entire waistline seam only when the dress does not have a sash. Before stitching on cording, at each side seam (A and A), pull out cord ⅝ inch and cut off. Then, restore shape of bias. Place raw edges of cording to meet raw edge of front of bodice. Using left side of zipper foot, stitch through same line of stitching on cording. Also, taper cording to nothing at side seams (A and A).

On outside of dress, stitch seam of sleeve and bodice ⅜ inch wide from sleeve down. Have armhole seam turned toward dress, because the corded edge of the facing may come too near the armhole and jam in machine. Press seam to one side or other so that well in seam will be eliminated when you turn to underside to complete stitching French seam. Trim seam to ⅛ inch, and to a point at edge of sleeve.

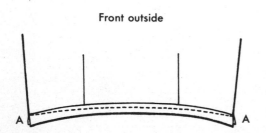

Front outside

A   A

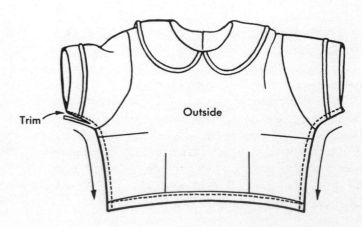

Trim

Outside

93

Stitch ¼-inch seam on underside from waist up; turn at edge of sleeve and stitch back for one inch.

The sleeves and bodice are always stitched in one operation on a child's dress, but never include any seams of any skirt at the same time.

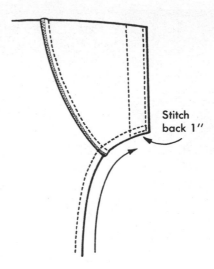

Stitch back 1″

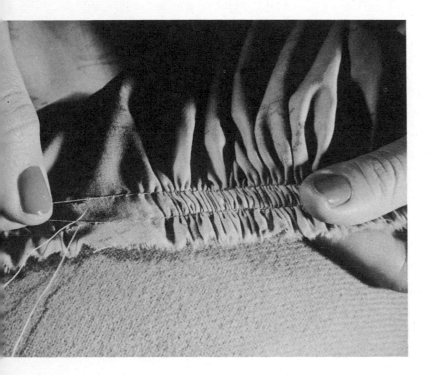

Pull up and tie gathering threads in each section of the skirt to fit bodice. Always pull the threads from the underside of the skirt and pull both of them in each section at one time. Do not make gathers on the placket opening at back.

With bodice side up, stitch bodice to skirt, using zipper foot; it is always needed when stitching cording. Turn right side of back facing to underside of skirt, and with bodice side up, sew facing through same line of stitching at waistline. Trim seam to ¼ inch, and round off to nothing at corners. Turn facings right side out, and beginning at end of facings, with gathers turned down, zigzag waistline seam at machine (see photograph, page 104). If you do not have a zigzag machine, make a second row of stitching ¼ inch from first row, and pink near second stitching. Then, you may also wish to overcast, by hand or by machine.

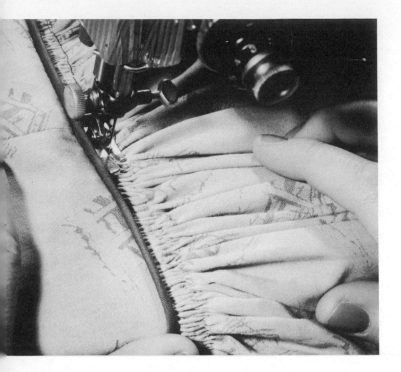

Using both hands and both wrists to hold gathers straight, and with a short stitch at machine, stitch through lower row of gathers for added strength.

We do not advise using elastic in the waistline seam of a child's dress.

Sash unit. Sash is always torn lengthwise with one edge on selvage, if possible. Make a strip long enough for both pieces of sash, but do not cut in half until stitching is completed. Staystitch ends (A) ¼ inch from edge.

Sash

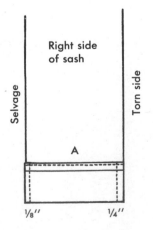

Turn and press raw edge on staystitching. Fold ends (A) to right side for 1¼ inches. On selvage side, stitch end ⅛ inch from edge, and on torn side ¼ inch from edge.

Turn right side out and press ends. Stitch sash, turning back selvage ⅛ inch, and raw edge ⅛ inch each time for a double turn.

When stitching around ends, follow sketch, so that topstitching can be continuous. Press sash.

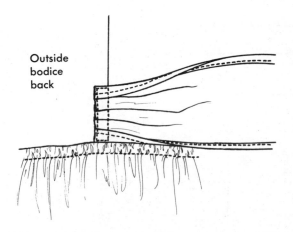

At half way mark, cut sash in two pieces. At cut end make 3 pleats in sash; also, fold back hem ¼ or ⅜ inch. Stitch ¼ inch from edge to hold in place.

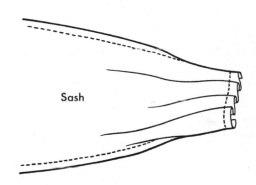

Sash

Place raw edge ¼ inch toward bodice back; stitch ¼ inch from edge in well of side seam, and make a second row on raw edge of sash. Turn to back of dress and press in place. If you would prefer a band in the front of the dress or belt in the back (or both), see second dress, pages 102–105.

Outside bodice back

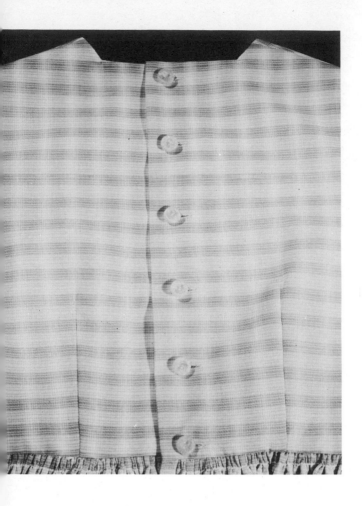

Completed dress unit. Make buttonholes in back of dress; the bottom buttonhole will be placed right at the top of the waistline seam. If the dress does not have a sash, use a hook and eye at the waistline, or a hook and a crocheted loop. Sew on buttons (page 84, *BMCC*).

We did not add any buttons for a trim below the collar on the front of the dress as commercially made ones so often have, because it is difficult to iron around them.

Final-press dress (Chapter 9, *BMCC*). For a French look on gathers, press flat for several inches below waistline.

Basic learnings for a dress with bodice underlined

Underlining is desirable in the bodice of a child's dress for several reasons. It gives more support to the bodice for adding various trims, and for the weight of the full skirt. The bodice is more pleasing in appearance in sheer fabrics, and it is more durable. Lastly, underlining gives a quality look to a child's dress.

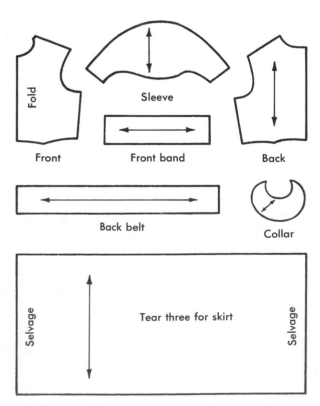

Skirt unit. Make precisely as in first dress, pages 84–87.

Bodice back unit. Cut bodice back and underlining ¼ inch wider than seamline for facing at edge A. In lighter weight fabrics, it is preferable to seam this edge rather than cut pieces double with a fold.

Staystitch neckline and shoulders just outside seamline in direction arrows indicate. Stitch darts; press to center back in bodice back, and to side edge in underlining.

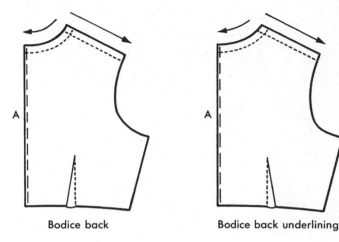

Bodice back   Bodice back underlining

With right sides of back and back underlining placed together, stitch ¼-inch seam at A edge. Tear a lengthwise reinforcement strip for buttons and buttonholes, 1½ inches wide and length of bodice back. Place ¼ inch from seamline, and stitch ⅛ inch from edges B, C, and D, and ⅝ inch from lower edge E. Trim strip away ⅝ inch up to E. Understitch underlining at A edge, beginning one inch from top.

With right sides together, stitch neckline to center back on seamline. Clip to stitching at center back, and trim seam to ¼ inch, rounding off seam allowance in corner. Turn and press, and press understitched edge down the back.

Even if a child's dress had smocking in the front and could not be underlined for support, the back could still be done as you have just learned.

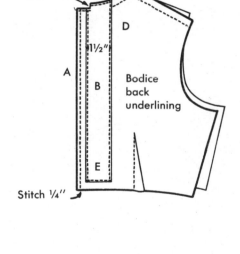

Bodice front unit. Staystitch neckline and shoulders just outside seamline and in direction arrows indicate on bodice front, and then on bodice front underlining. Stitch darts; press toward center front in bodice front, and toward side edge in underlining.

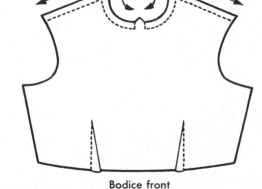

Bodice front

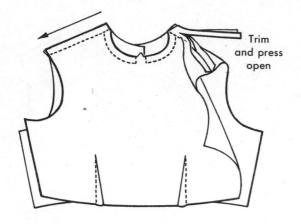

Trim
and press
open

Bodice unit. Stitch shoulder seams in direction arrows indicate on bodice and then on underlining. Trim to ⅜ inch and press open.

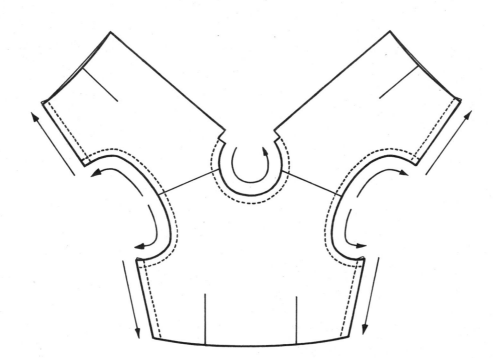

The inside of the bodice will face the inside of the underlining. Since the neckline has already been staystitched, you may disregard direction of grain, but staystitch neckline again *through both layers* just outside seamline from center back to center back.

Staystitch armholes and side edges together just outside seamline and in direction arrows indicate.

Collar unit. With shortest stitch machine will make, and with right sides together, stitch under collar to top collar along outer curved edge. Trim seam to a *very scant* ⅛ inch. Do not try to understitch this collar, because it is too small a circle.

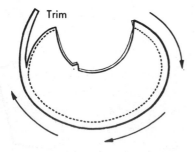

Trim

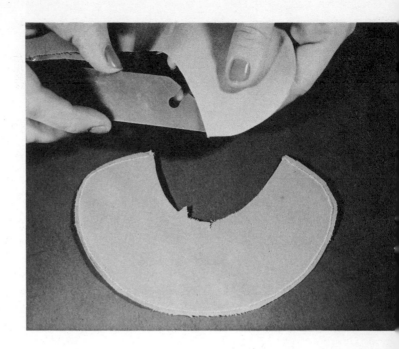

Turn right side out, and use a point former to round out edge of collar before pressing. Keep seam under point former when working around edge of collar; press collar.

At neckline, stitch together top and under collar, just outside seamline. Key the two collars precisely and overstitch about 1 inch to hold together, so they will not slip or spread when attached to center front of bodice. If a trim is being used on collar that will extend to the neckline, put on trim at this time.

Place collar in position at center front and center back, and with dress side up, stitch collar to dress with ⅝-inch seam allowance (see sketch on page 90). Clip curve of dress and collar frequently up to seamline to straighten curve for ease in stitching. Trim entire neckline seam to scant ¼ inch.

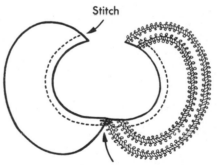

Stitch

Stitch collars together

If the colors will match, you may use commercial cotton or rayon bias, or cut a piece one inch wide. With dress side up, sew on bias through previous stitching. Do not stretch bias; allow it to extend freely ¼ inch beyond center back. If necessary, trim width of bias, so it will just turn in a narrow amount and fit over scant ¼-inch seam. Commercial bias is usually trimmed to fold line of bias. Turn back raw edges at center back, and stitch bias flat to dress around neckline. We have eliminated facings at the neckline for ease in ironing.

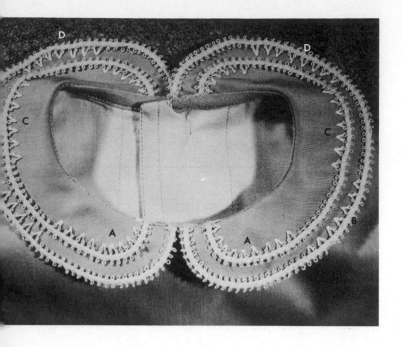

This is a photograph of the completed collar. The commercial trimming was tacked in place blindly. Then, an embroidery stitch with a contrasting color was used in a staggered fashion (A, B, C, and D) for a decorative effect. The same idea was repeated on the skirt.

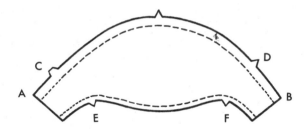

Sleeve unit. Staystitch cap of sleeve (A to B) exactly on seamline, changing to longer stitch at machine between notches (C to D). Staystitch lower edge of sleeve scant ⅜ inch from edge, changing to longer stitch at machine between notches (E to F).

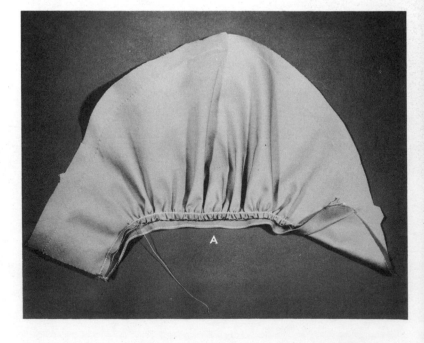

To bind lower edge of sleeve, cut a bias strip 1½ inches wide and the length of a comfortable arm measurement for the child plus two seam allowances underarm. On the inside of the sleeve, clip the first long stitch at either E or F, and draw up sleeve to fit bias strip. With sleeve side up, stitch sleeve to bias, sewing one thread inside previous stitching. Trim seam to perfect ¼ inch. Turn over bias and press in place. Turn in raw edge of bias (A) and edge-stitch fold of it. The bias must be wide enough for the folded edge to extend about ⅛ inch beyond the stitching line on underside of the dress.

To prepare sleeve to be set in armhole, clip the first long stitch at the notch on the right side of the inside of the sleeve. Draw up sleeve to fit armhole and fasten thread. Distribute gathers, and with sleeve side up, stitch sleeve to armhole of dress. Sew one thread inside previous stitching (see photograph, page 63). Trim seam to ⅜ inch, and with gathers turned down, zigzag near edge of seam at machine (see photograph, page 70). If you do not have a zigzag machine, make a second row of stitching ¼ inch from first row, and pink near second edge. Then, you may also wish to overcast, either by hand or by machine.

With fabrics like organdy, finish the armhole seams as described on page 93 for the first dress.

Back belt and front band unit. The belt for the back of the dress is always torn lengthwise. Make a strip long enough for both parts of the belt, but do not cut in half until stitching is completed. If you would prefer a sash in the back or cording in the front of the dress (or both), see first dress (pages 93–97). Tear a lengthwise strip 3 inches wide and the width of lower edge of bodice back plus approximately 6 inches extra.

With right sides together, stitch ends and length of torn edges with ¼-inch seam; use shortest stitches at machine one inch each side of halfway mark (A). Trim corners, and at halfway mark, cut belt in two pieces. Press open seam on a dowel or ruler; turn belt right side out and press.

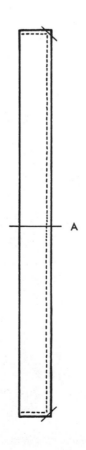

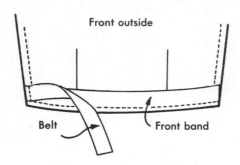

Front outside

Belt

Front band

The band for the front of the dress is always torn lengthwise. Tear it 3¼ inches wide (to finish 1¼ inches, same as belt, plus ⅜-inch seam allowance) and the width of lower edge of bodice front. With right sides out, fold in half and stitch raw edges together ⅜ inch from edge. Place raw edge of band to meet raw edge of bodice front, and stitch through previous stitching on band ⅜ inch from edge.

Place belts on sides of bodice front with folded edge at fold of band; lower edge of belts should key with ⅜-inch seamline. Stitch to dress just outside seamline.

Complete bodice unit. With right sides together, stitch seam of sleeve and bodice from sleeve down, including bias at lower edge of sleeve. The sleeves and bodice are always stitched in one operation on a child's dress, but never include any seams of any skirt at the same time.

Trim seam to ⅜ inch, and zigzag near edge of seam at machine.

Begin below bias on sleeve edge, and turn armhole seam toward sleeve. If you do not have a zigzag machine, make a second row of stitching ¼ inch from first row, and pink near second edge. Then, you may also wish to overcast.

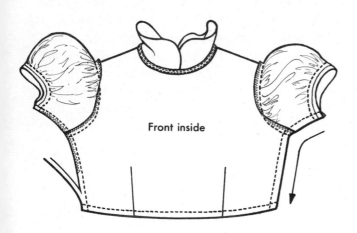

Front inside

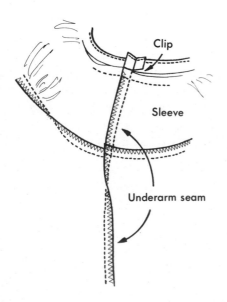

Clip

Sleeve

Underarm seam

Clip bias at end of zigzag stitch so it can lie open.

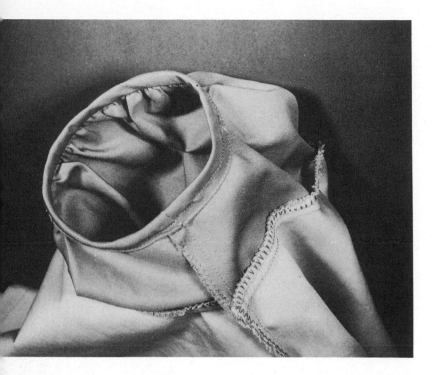

Restore turn of bias at seamline, and with sleeve side up, stitch bias in place through well at seamline.

Pull up and tie gathering threads in each section of the skirt to fit bodice. Always pull the threads from the underside of the skirt and pull both of them in each section at one time (see photograph, page 94). Do not make gathers on the placket opening at back.

With skirt side up, stitch bodice to skirt, sewing through first row of gathers ⅜ inch from seamline. With gathers turned down, zigzag waistline seam at machine. If you do not have a zigzag machine, make a second row of stitching ¼ inch from first row, and pink near second stitching. Then, you may also wish to overcast.

At back opening, turn in waistline seam at an angle (A), and bar tack at machine.

Using both hands and both wrists to hold gathers straight, and with a short stitch at the machine, stitch through lower row of gathers for added strength (see photograph, page 95).

We do not advise using elastic in the waistline seam of a child's dress.

Completed dress unit. Make buttonholes in the back of the dress and in the belt. The bottom button-hole on the dress will be placed right at the top of the waistline seam (see photograph, page 97). If the dress does not have a belt, use a hook and eye at the waist-line, or a hook and a crocheted loop. Sew buttons on dress and belt (pages 84–85, *BMCC*).

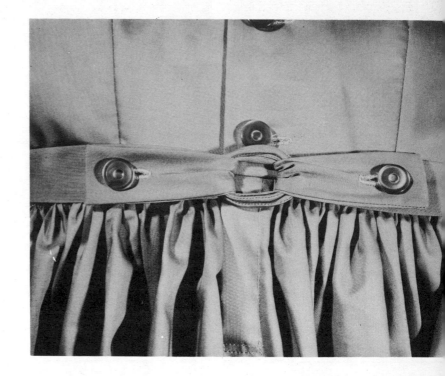

The belt may overlap about 3 inches and be fastened with two buttons and buttonholes.

The ends of the one in the photograph go through a triple ring of bias tubing and fasten back on the belt with buttons and buttonholes.

If trim is used on skirt, a seam may be opened to insert raw edge of trim; stitch seam again. However, trim may be applied before skirt is put together at last seam.

Final-press dress (Chapter 9, *BMCC*). For a French look on gathers, press flat for several inches below waistline.

Additional suggestions for quality looking dresses.

Every little girl should have a fine broadcloth smocked dress. The hem of this dress features the technique for lengthening shown on page 109. The embroidery is done on the top of the hem, and then the hem is sewn by hand to the dress.

The top of this organdy pinafore is underlined as shown on pages 97–99. The hem features the technique for lengthening shown on page 109. The trimming is sewn fast to the top of the hem, and then the hem is sewn by hand to the dress.

The facings for this dress are cut like those for the blouse on page 35, *BMCC*. This simplifies finishing the neckline and reinforcing the armholes.

Sew together shoulder seams of front and back bodice. Then, stitch bias cording to neckline of bodice just as you learned to do on the top collar, page 89.

Sew shoulder seams of facings, and apply facings to bodice just as you learned to do on under collar, page 90.

Staystitch armhole edges together, and proceed with techniques for completing dress that you have learned in this chapter.

Children's dresses

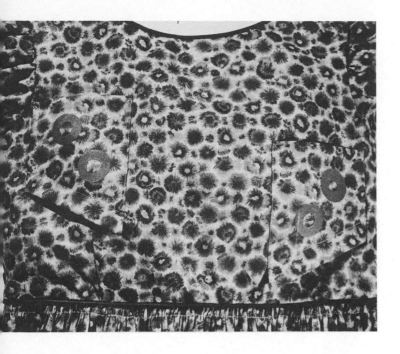

The pockets on this dress are trimmed with Bondex, which comes in assorted colors and can be cut into desirable shapes and ironed on garments.

The bias on the sleeves, bodice, waistline, and skirt of this dress were done as in the photograph shown on page 225.

The two parts of the sash were made on the bias, and the bias band at the front was topstitched over them (A and A).

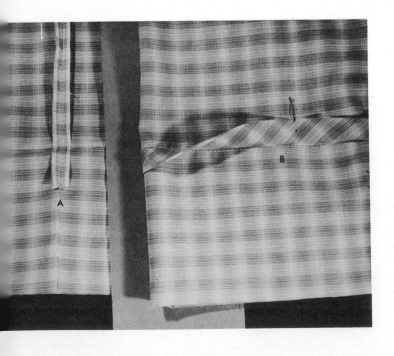

By clipping seam (A) and turning it to right side of dress under hem, the hem can be lowered on right side of dress (B), and creases or the worn hem edge will not show.

This is not possible on ready-to-wear, but you can stitch a narrow tuck at crease or worn hem edge before hemming dress. Another possibility is to add a trim such as braid over the crease or worn hem edge.

Short shorts, pedal pushers, bermuda shorts, and slacks for women and girls

Most likely, there will not be another chapter in this book that will mean more to the American girl and woman than this one. It isn't that they cannot buy short shorts, pedal pushers, bermuda shorts, and slacks, but that they cannot buy them to fit, or even have them altered to fit. We find more ill-fitting pants on the American woman than good-fitting ones. *A good fit starts in the cutting for the individual figure.*

Many of the techniques in fitting and cutting-to-fit will be similar to those for a skirt on pages 65–70, *Bishop Method of Clothing Construction.*° Also, you will need to use the skirt key grain lines on page 49, *BMCC.* Work with an inexpensive piece of fabric to perfect the fit of pants for your individual figure; then, you can apply your learnings to any of them you make in the future.

Whether you are tall or short, take the hip measurement over the largest part of the hips to determine pattern size. It is important for slacks to fit as closely as possible at the hip because of the shaping of the crotch. Proportioned slacks fit as far as height is concerned.

The following cutting-to-fit suggestions are the most common ones needed for the majority of women on any length of pants.

° Hereafter referred to as *BMCC.*

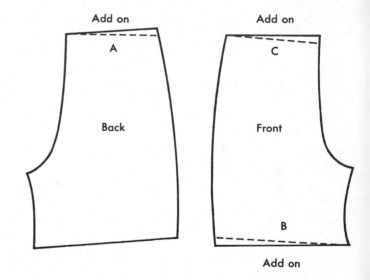

Cutting-to-fit

At A, ½ inch to as much as one inch can be added at the side back for the figure that has fullness at the side hipline, and side seams swing toward front.

At B, add a corresponding amount, so that side seams will match with perfection. Do not add as much as you did at the back, however, if B edge will fall below straight of grain at side edge. In that case, ease difference on back to front of shorts.

At C, approximately ½ inch can be added (not more than will make it straight of grain at center front) for the person who has fullness in the front of her figure.

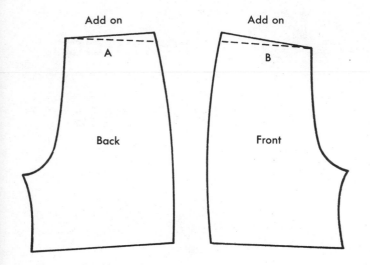

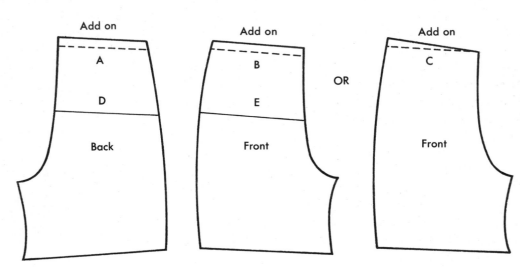

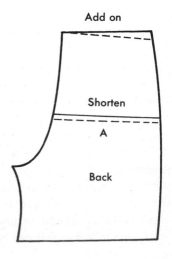

If the figure has fullness at side hipline on both the front and back (see former sketch for back only), ½ inch to as much as one inch can be added on waist edge at side seam (A and B), tapering to nothing at center front and center back. If the grain tells you to do so, you may add a little extra at the center front, also.

If the figure is full across the back of the hips, or if the figure requires a longer crotch, add needed amount of fabric on waist edge of pants back (A). On the front, needed amount of fabric may either be added all the way across waist edge (B) or tapered from side seam to nothing at center front (C). It may be added at D and E (line indicated on pattern) instead of A and B, if the figure is wide in the hip.

If the figure is flat in the hips, take out necessary amount by making an even fold (A) in pattern halfway down crotch seam. Then, add a corresponding amount at waist edge tapering to nothing at center back, so that side seams will match with perfection. See sketch on page 112 also.

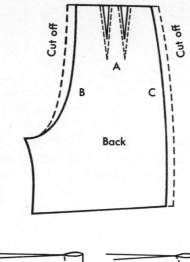

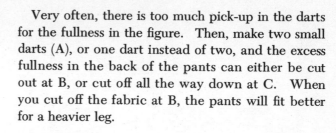

Very often, there is too much pick-up in the darts for the fullness in the figure. Then, make two small darts (A), or one dart instead of two, and the excess fullness in the back of the pants can either be cut out at B, or cut off all the way down at C. When you cut off the fabric at B, the pants will fit better for a heavier leg.

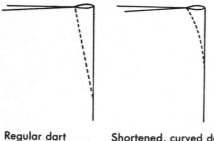

Regular dart   **Shortened, curved dart**

Frequently, on the front of the pants, the long, straight dart stitches out fabric that is needed for ease in the fit of the figure. Then, the shortened, curved dart (A) is preferred, because it will give more fabric where it is needed, or one or two tucks (B) may be folded where ease is needed, instead of using the dart.

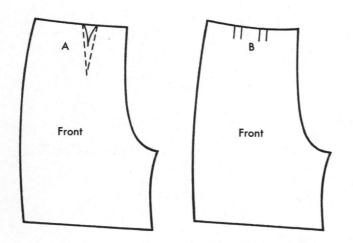

When cutting pants from plaid or striped fabric, the design will be improved on the back if line is made straighter than pattern at A. Add corresponding amount at B to front and back. If the figure is heavy in the thighs, add fabric all the way down at B as shown in the second sketch. This alteration will also improve the fit of pants for figures that are flat in the back (see sketch on page 111).

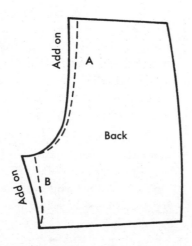

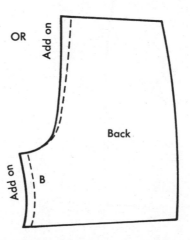

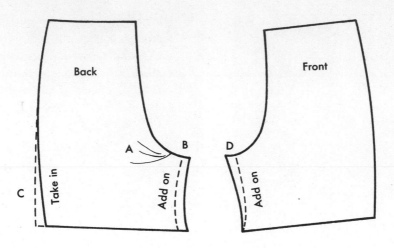

When pants sink in at crotch (A), and pull or wrinkle at the same place across the front, it is because they need more room on the inside of the leg (B).

If the pants look wide across the back when necessary amount has been added at B, taper side seams (C).

Add corresponding amount at D, tapering to nothing at lower edge.

It may be necessary to add all the way down the front at D. If so, the side seams of the front may be tapered, also.

This is a valued alteration for the person who has a heavy inside leg.

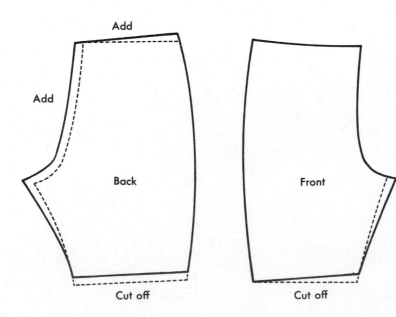

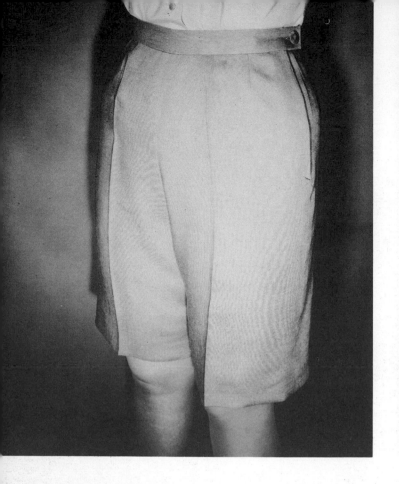

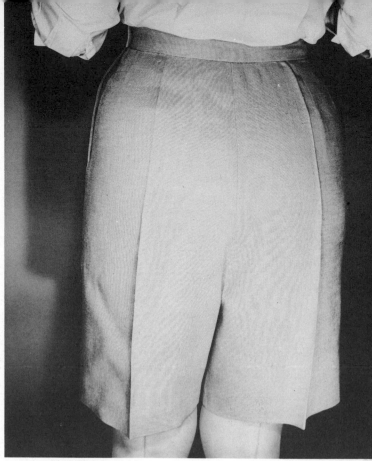

The bermuda shorts in the photographs have been fitted to grain perfection, using the two all-important key grain lines on the sketch.

These alterations were made on a size 14 pattern to the bermuda shorts in the above photographs.

Key grain lines

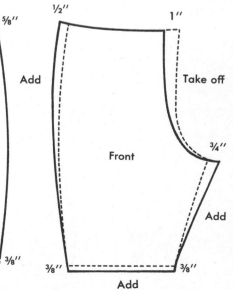

Short shorts

The short shorts may be underlined, if desired, with the same techniques shown with bermuda shorts, pages 128–133. If they are not underlined, make as follows:

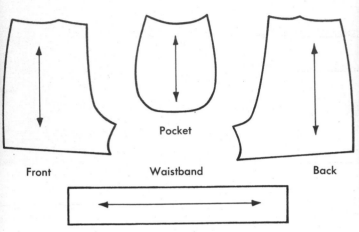

Front   Waistband   Back

Back unit.   Staystitch waist edge and right back of shorts from notch up just outside seamline and in the direction shown with arrows on sketch.

Staystitch left back of shorts from notch up and lower edges scant ¼ inch from edge and in the direction shown with arrows on sketch.

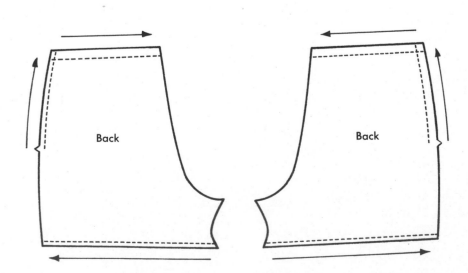

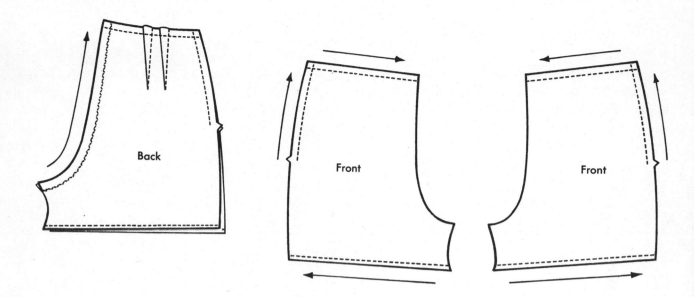

Sew seam at crotch in direction shown. Considerable strain is placed on this seam; strengthen it by stitching with machine set one point or even less to zigzag. If you do not have a zigzag machine, sew with shortest stitch at machine.

Set the machine more than one point to zigzag and stitch both sides of seam ¼ inch from edge. Stretch seam in curve only. If you do not have a zigzag machine, staystitch seam, and stretch in curved area, also. Press open seam.

Stitch darts (page 27, *BMCC*) and press toward center back.

Front unit. Staystitch waist edge and sides of shorts from notch up just outside the seamline and in direction shown with arrows on sketch. Staystitch lower edges scant ¼ inch from edge and in the direction shown with arrows on the sketch.

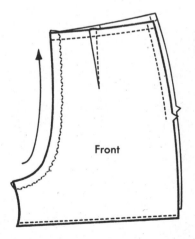

Sew seam at crotch in direction shown. Considerable strain is placed on this seam; strengthen it by stitching with machine set one point or even less to zigzag. If you do not have a zigzag machine, sew with shortest stitch at machine.

Set the machine more than one point to zigzag and stitch both sides of seam ¼ inch from edge. Stretch seam in curve only. If you do not have a zigzag machine, staystitch seam and stretch in curved area, also. Press open seam.

Stitch darts (page 27, *BMCC*) and press toward center front.

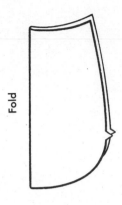

Fold

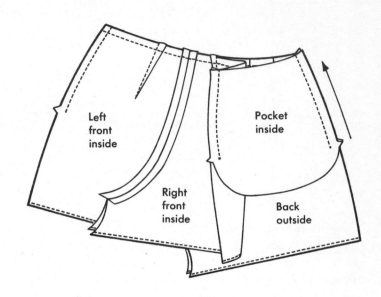

Left front inside

Pocket inside

Right front inside

Back outside

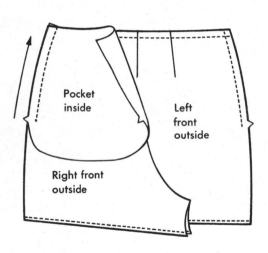

Pocket inside

Left front outside

Right front outside

Pocket unit.   Cut pocket on the fold for a smooth line in shorts.   Use one pocket on the right front of shorts.

With right sides together, stitch front pocket piece to front of shorts from notch to the waist edge. Lockstitch securely and precisely at notch or other pattern marking for pocket opening.

With right sides together, stitch back pocket piece to back of shorts from notch to the waist edge.   Lockstitch securely and precisely at notch or other pattern marking for pocket opening.

117

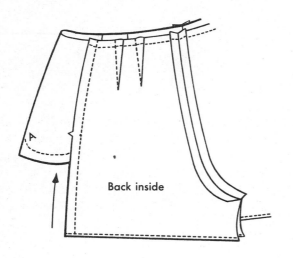

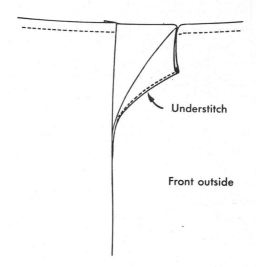

Join front and back unit of shorts by stitching side seam from lower edge up to pocket opening; this connects the stitching line of the pocket pieces.

Check the cut of shorts for direction to sew side seams. If the style is tapered at lower edge, always key lockstitching on pocket pieces and sew from lockstitching down to lower edge.

Then, place the shorts away from the pocket pieces, and starting exactly at lockstitching, close pocket pieces from notch to fold of pocket, rounding off stitching in corner to keep lint out (A).

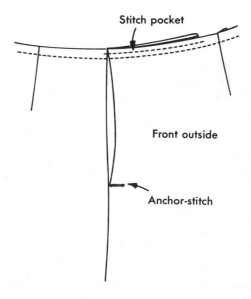

Trim the front side seam to ¼ inch, and understitch edge of front pocket piece length of pocket opening.

Place pocket in position on front of shorts, matching upper edges with waist edge. Stitch in place.

Press the back side seam toward back of shorts, and below the pocket opening; front pocket seam is pressed toward the front. Anchor-stitch at lower edge of opening.

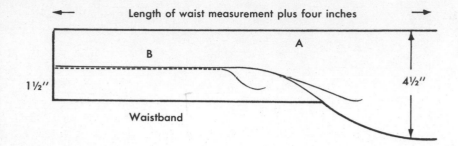

Length of waist measurement plus four inches

B

A

1½″

4½″

Waistband

Zipper unit. On left side of shorts, stitch side seam and insert placket zipper precisely like series of learnings on pp. 40–41, *BMCC*.

Waistband unit. The waistband, to finish 1¼ inches wide, is made from a lengthwise torn strip 4½ inches wide and about 4 inches longer than individual waist measurement. With a crosswise-ribbed fabric such as bengaline, a crosswise strip would be used for the band.

With automatic hem gauge, make a lengthwise fold to the inside 1½ inches deep, and press. Stitch close to raw edge (B) of folded section; this forms the band's own interfacing.

To attach the waistband to the shorts, place the right side of band (edge A) to the wrong side of shorts. Pin first at center front (R), allowing enough additional band to extend to front placket opening plus 1¼ inches extra (S).

Measure one-half of waist measurement and allow that much band to center back of shorts (T). Pin in place. The size is accurate from center front to center back. For example, if the waist measures 28 inches, the amount of band from center front to center back would be 14 inches.

Next, fold the shorts from center front to center back to measure one-half of them, and mark with a pin (U). Repeat with the band (U) and bring together the two pins. The pattern manufacturer has given tolerance on shorts to allow them to fall softly over normal body curves, and this will aid in distributing it evenly; pin shorts to band at intervals. The halfway mark on the shorts (U) will not be the side seam, because the front of the shorts is always cut wider than the back of the shorts.

To measure the amount of band needed for the second half of the back of the shorts, fold loose end of the band carefully back to the side seam. Mark with a pin on the band the exact seamline. Place the pin on seamline of zipper edge of other half of back of shorts (V). Distribute ease evenly and pin band to shorts at intervals. Repeat for second half of band front and front of shorts, and pin at S.

Sew band to shorts on seamline with band side up. It is easier to control the ease on shorts when it is on the underside, and you will be able to follow a check, stripe, or ridge on band (when fabric has such a design) for perfection in stitching and in appearance on outside of band.

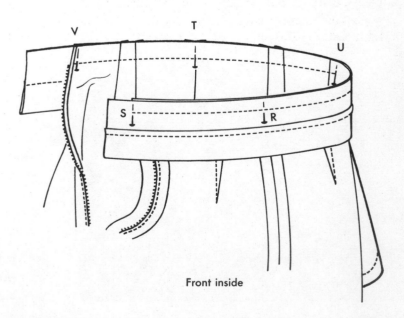

V   T   U

S   R

Front inside

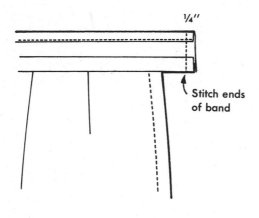

¼″

← Stitch ends
of band

Press seam up. To stitch ends of band, fold right sides together and stitch ¼-inch seam. The belt extends one inch on the front and laps under approximately two inches on the back.

Trim seam to ¼ inch and round off seam allowance at top corners. Turn band to outside of shorts and pin in place.

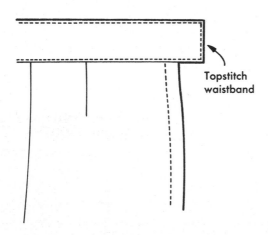

Topstitch
waistband

The band is already turned in with perfection and is ready to topstitch on lower edge over seamline on shorts. Continue to topstitch all around band.

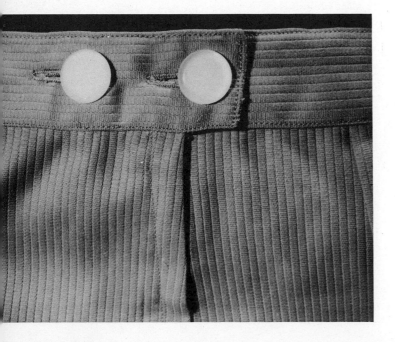

Finish ends of band with two buttons and two machine buttonholes. To sew on buttons, see pp. 84 and 85, *BMCC*.

Hem unit. Using automatic hem gauge, turn under raw edge ¼ inch and press. Then, turn up hem 1¼ inches and press. Repeat for second half of shorts.

Stitch inside leg seams from the top down and press open.

Hem shorts at machine as in bottom photograph, page 46, *BMCC*.

Final-press shorts (Chapter 9, *BMCC*).

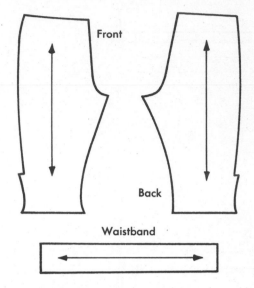

Front

Back

Waistband

Pedal pushers

The pedal pushers may be underlined, if desired, with the same techniques shown with bermuda shorts, pages 128–133. If they are not underlined, make as follows:

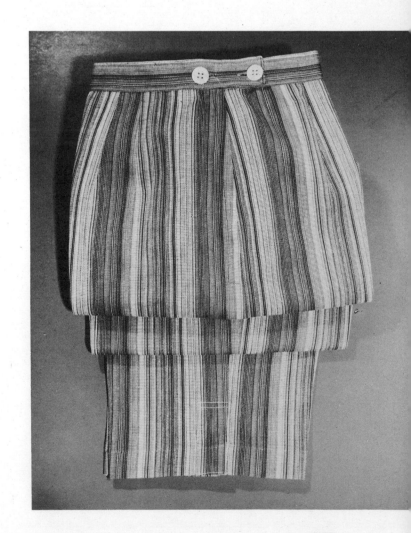

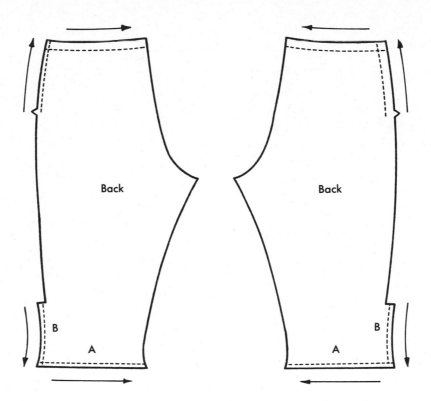

Back unit. Staystitch waist edge and right back of pedal pushers from notch up just outside seamline, and in the direction shown with arrows on sketch.

Staystitch left back of pedal pushers from notch up and edges A and B scant ¼ inch from edge and in the direction shown with arrows on sketch. If pattern for pedal pushers has a fitted facing at lower edge, see techniques at end of this section.

Sew seam at crotch in direction shown. Considerable strain is placed on this seam; strengthen it by stitching with machine set one point or even less to zigzag. If you do not have a zigzag machine, sew with shortest stitch at machine.

Set the machine more than one point to zigzag and stitch both sides of seam ¼ inch from edge. Stretch seam in curve only. If you do not have a zigzag machine, staystitch seam and stretch in curved area, also. Press open seam.

Stitch darts (page 27, *BMCC*) and press toward center back.

Using automatic hem gauge, turn under raw edge at A ¼ inch and press. Then, turn up hem 1¼ inches and press (C).

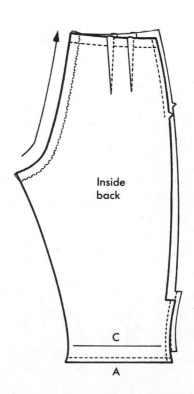

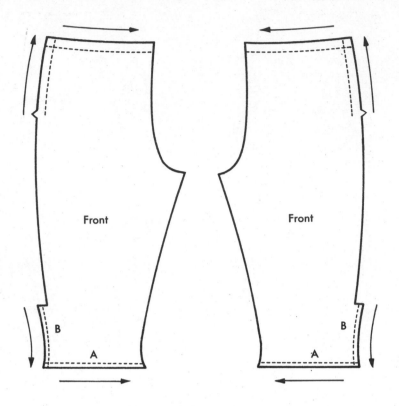

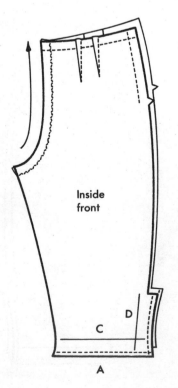

Front unit. Staystitch waist edge and sides of pedal pushers from notch up just outside the seamline and in direction shown with arrows on sketch.

Staystitch edges A and B scant ¼ inch from edge and in the direction shown with arrows on sketch.

Sew seam at crotch in direction shown. Considerable strain is placed on this seam; strengthen it by stitching with machine set one point or even less to zigzag. If you do not have a zigzag machine, sew with shortest stitch at machine.

Set the machine more than one point to zigzag and stitch both sides of seam ¼ inch from edge. Stretch seam in curve only. If you do not have a zigzag machine, staystitch seam and stretch in curved area, also. Press open seam.

Stitch darts (page 27, *BMCC*) and press toward center front.

Using automatic hem gauge, turn under raw edge at A ¼ inch and press. Then, turn up hem 1¼ inches and press (C). Turn extension to inside on pattern fold line (D) and press.

Complete unit. If a pocket is desired on the right side of pedal pushers, follow the techniques for pocket in short shorts described on pages 117–118.

Sew side seams from waist edge to pattern marking for opening (A). Edges are staystitched above notches, thus making it possible to stitch off grain in this short area. Leave left side open above notch for placket zipper. On back of pedal pushers, clip to stitching at pattern marking for opening (A). Next, sew inside leg seams from top down. Pink seams with pinking shears, if necessary, and press open. Restore hem line at lower edges in pressing.

Hem pedal pushers at machine as in bottom photograph, page 46, *BMCC*.

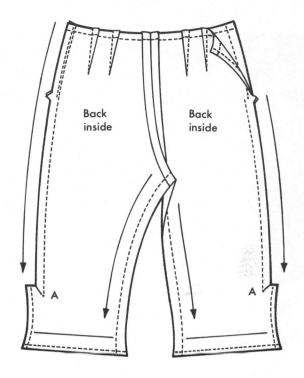

Zipper unit. On left side of pedal pushers, insert placket zipper precisely like series of learnings on pages 40–41, *BMCC*.

Side openings at lower edge. 1. Back. Turn in raw edge on ¼-inch staystitching and turn again to make ½ inch wide hem. Press. Stitch near edge of hem (A) at machine, and form a triangle (B) at lower edge as shown.

2. Front. The front of the opening is already pressed on pattern marking line (C). Turn in raw edge on ¼-inch staystitching and stitch across bottom (D), up the hem near folded edge (E), and back and forth through front and back (F) for reinforcement.

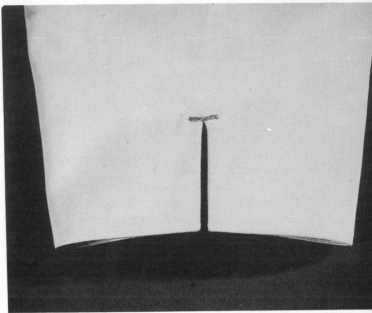

This photograph shows the completed opening from the topside.

Waistband unit. The waistband is put on precisely like the series of learnings on pages 119–120 for short shorts.

Final-press pedal pushers (Chapter 9, *BMCC*).

Fitted facing at lower edge. If the pattern has a fitted facing at lower edge of pedal pushers, staystitch lower edges of pedal pushers (A), front and back, and facings (B) just outside seamline. Staystitch top of facings (C), front and back, scant ¼ inch from edge.

Clean-finish upper edges of facings (C). To clean-finish, turn under on ¼-inch staystitching line and stitch close to folded edge.

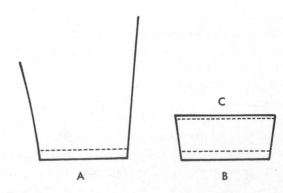

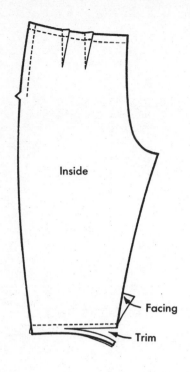

Inside

Facing

Trim

Right sides together and pedal pushers turned up, sew facings to lower edges, front and back. Trim seam to scant ¼ inch, and understitch edge, beginning and ending one inch from edges. Understitching is a row of machine stitching along the edge of the facing that catches the two trimmed seams to the facing.

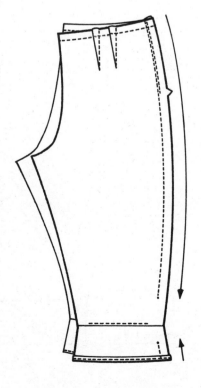

Sew side seams of pedal pushers from waist edge to pattern marking for opening. Edges are stay-stitched above notches, thus making it possible to stitch off grain in this short area. Leave left side open above notch for placket zipper. Sew side seams of facing down to pattern marking for opening.

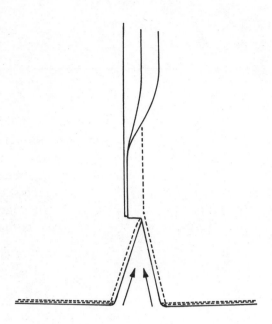

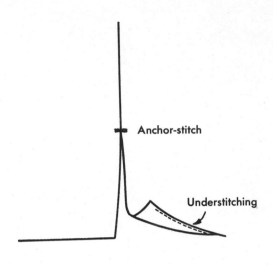

Anchor-stitch edge of opening.

Right sides together, and beginning at lower edge of pedal pushers, close point of opening down to pattern marking for opening. Repeat for second half of point. Round off seam allowance in corners and trim seams to ¼ inch. Trim remainder of facing seam to ¼ inch.

Turn right side out and press. Press open side seam and facing seam.

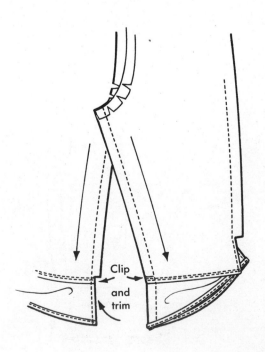

Open out facing at inside leg seam and sew inside leg seam and facing from crotch down. Press open seam. Clip up to stitching at seamline at fold of facing, and trim facing seam to ¼ inch. Turn back facing and press.

Stitch hem at machine same as other pedal pushers and short shorts.

Lined bermuda shorts

All of the quality-looking bermuda shorts are under-lined. However, they can be made without under-lining, following the techniques for short shorts at the beginning of this chapter, just as short shorts, pedal pushers, and slacks could be underlined, using the techniques from this section.

Popular underlinings are cotton broadcloth, cotton percale, rayon twill, and super soft siri.

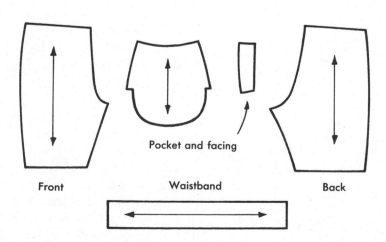

Front   Waistband   Back

Pocket and facing

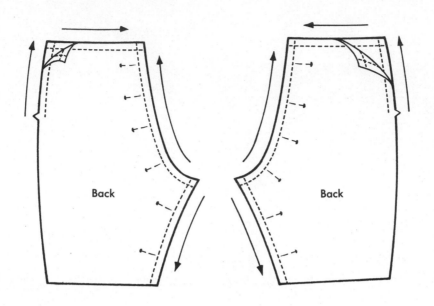

Back　　　　　　　　Back

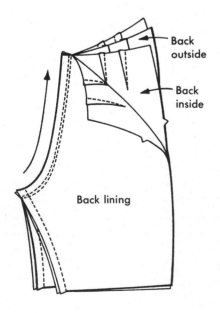

Back outside

Back inside

Back lining

Back unit. Place underlining fabric to inside of shorts fabric. Pin together at crotch seam and inside leg seam. *Separately*, staystitch shorts and underlining at waist edge and sides from notch up, just outside the seamline, and in direction shown with arrows on sketch.

Together, staystitch crotch seam and inside leg seam just outside seamline, and in direction shown. On one half, the underlining will be on top; on other half, shorts will be on top. With loosely woven fabric for shorts, it may be difficult to sew with shorts on top and underlining below. Then, guard grain very cautiously and sew against it—with underlining on top.

Sew seam at crotch in direction shown. Considerable strain is placed on this seam; strengthen it by stitching with machine set one point or even less to zigzag. If you do not have a zigzag machine, sew with shortest stitch at machine.

Set the machine more than one point to zigzag and stitch both sides of seam ¼ inch from edge. Stretch seam in curve only. If you do not have a zigzag machine, staystitch seam and stretch in curved area, also. Press open seam.

Separately, stitch darts (page 27, *BMCC*), so that underside of darts in shorts and in underlining will face each other. Press darts in shorts to center front and in underlining to side edge.

Sometimes, the darts are not stitched in underlining and are just folded in place.

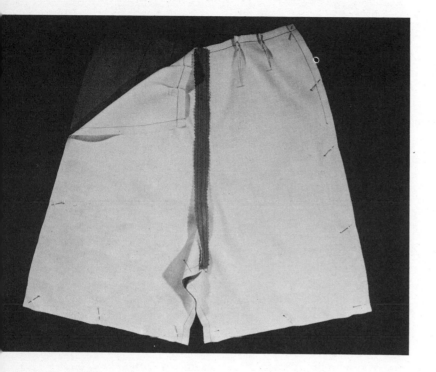

Front Unit. The front unit is made following the techniques above for the back unit.

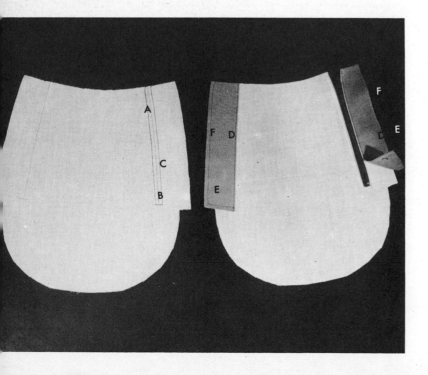

Pocket unit. On pattern marking line (A) for zipper insertion in left pocket, staystitch for length of zipper and tape at top. Cross over at lower edge of staystitching (B) and make a second row of stitching (C) ⅜ inch from first row. Slash between rows of staystitching and into corners at lower edge. Press under raw edges on staystitching. With zipper foot, stitch pocket piece close to metal of zipper.

Staystitch D edges of pocket facings scant ¼ inch from edge. Press under raw edge on staystitching. With the zipper foot, stitch D edge of pocket facing close to metal of zipper. With regular presser foot, stitch D edge in place on other half of pocket, and stitch E and F raw edges to pocket, ¼ inch from edge.

Make a second pocket for right side, if desired.

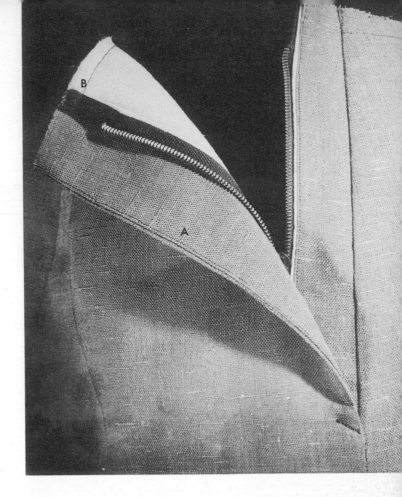

Insert the pocket with the same series of steps shown with short shorts on pages 117–118.

On front and back pocket seam, trim muslin to seamline. Stagger remaining edges of front seam in trimming, making upper one ¼ inch and under one ⅛ inch. Understitch edge of front pocket (A) length of pocket opening. Open the zipper, and place remainder of pocket in place on front of shorts, matching upper edges with waist edge (B). Stitch in place.

Press open the side seam below the zipper. Anchor-stitch at lower edges of pocket opening. If right pocket is used, anchor-stitch one inch down from upper edge, also.

Underlining unit. Going from top down, stitch side seams of underlining so that raw edges of seams will face shorts. Press open seams.

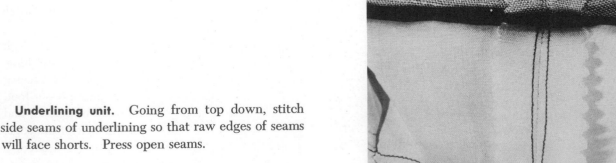

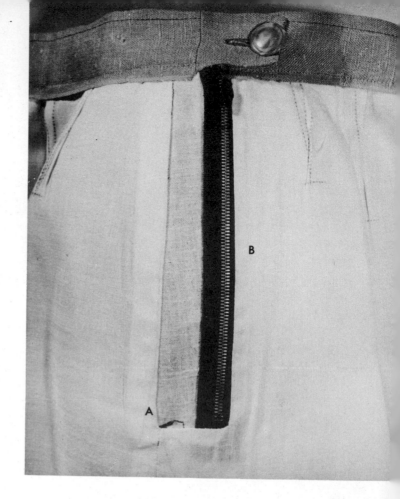

On left side, leave open seam to bottom of metal part of zipper. At A, clip to lockstitching on back seam allowance; lay raw edge of underlining to meet raw edge of shorts' side seam. Zigzag raw edges together at machine, or slipstitch by hand.

Keep zipper closed. Lay underlining in place on front of shorts. Chalk mark or pin mark where you can feel the front edge of metal zipper on underlining (B). Staystitch this mark, and also straight across lower edge of metal zipper to lockstitching on side seam of underlining. Beginning above lockstitching, cut underlining at an angle into staystitching at corner. Trim away excess fabric, leaving ½-inch seam allowance to turn under to staystitching. Whipstitch underlining in place along stitching line on front placket edge, and on lower edge over to lockstitching at seamline.

With underlining on top, stitch together raw edges of shorts and underlining at the waistline, just outside the seamline.

Waistband unit. If bermuda shorts are being made of a lightweight fabric such as cotton cord, waistband will be made and put on precisely like the short shorts pages 119–120). However, if shorts are being made of linen, lightweight wool, etc., use the following technique.

The waistband (to finish 1¼ inches wide) is made from a lengthwise-torn strip 3½ inches wide and about 4 inches longer than the individual waist measurement. Tear with one edge along the selvage. With a crosswise-ribbed fabric such as bengaline, it is a crosswise strip for the band.

Interfacing (underlining fabric is usually satisfactory) is also a lengthwise-torn strip 2 inches wide.

Place interfacing ¼ inch from torn edge of band and stitch in place near edge of interfacing. Then, stitch second edge of interfacing to band. It is not necessary to press band before stitching second edge, as we learned to do on page 170, *BMCC*, for heavier fabrics in tailoring a suit. Try band around waistline, and hold with firmness one likes in fit of band, to determine size needed.

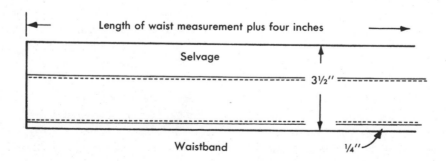

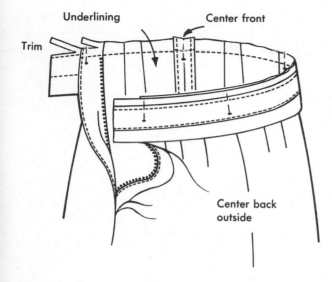

Trim · Underlining · Center front · Center back outside

Selvage will fall in place ⅜ inch below seamline on back of shorts; pin to hold in place. (See photo page 132.) With shorts right side up, stitch in well on seamline to hold band in place.

Final-press the band. Finish ends with two large metal hooks and eyes, a double set of large snaps, or two buttons and buttonholes.

To attach waistband to shorts, open zipper and place torn edge of band to meet waist edge of shorts, right sides together. Pin first at center front, allowing enough additional band to extend to edge of left front pocket opening, plus a seam allowance extra.

Proceed with pinning on band exactly as you learned to do with the skirt on page 43, *BMCC*, or short shorts, pages 119–120.

Sew band to shorts on seamline with band side up. It is easier to control the ease on shorts when it is on the underside, and you will be able to follow a check, stripe, or ridge on band (when fabric has such a design) for perfection in stitching and in appearance on outside of band. Press seam up and trim band seam allowance to ¼ inch and shorts to ⅜ inch.

To finish ends of band, fold right sides together. Bring selvage edge ⅜ inch over seamline on front of band. On back of band, stitch band parallel to edge of pocket opening, and on front of band to extend 1½ inches. Trim seams on both ends to ¼ inch and round off corners. Turn ends right side out and press.

Hem unit. Trim seam in hem of shorts to ¼ inch, going a little beyond fold of hem. With underlining side up, stitch underlining and shorts together 1¼ inches from raw edge (A). Using hem gauge, turn up 1½ inches and press lower edge of shorts and underlining together. Previous stitching will be ¼ inch from fold of hem; trim away underlining to stitching.

Staystitch or zigzag shorts ¼ inch from raw edge of hem (B). Pink with pinking shears, if necessary.

Stitch inside leg seams from top down, and press open. Trim seam in hem of shorts to ¼ inch, going a little beyond fold of hem.

Stitch hem to underlining by hand, using pick stitch on underside of hem (page 82, *BMCC*).

Final-press shorts, Chapter 9, *BMCC*.

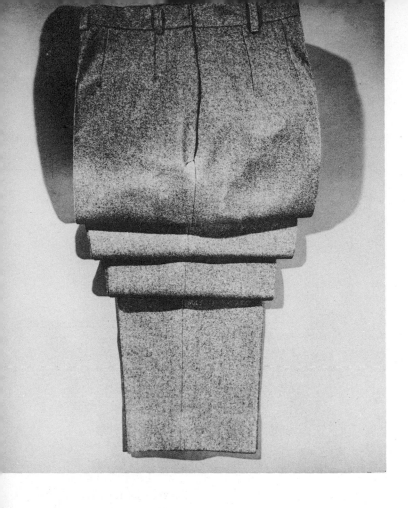

Slacks

We are not showing the techniques from beginning to end for making the slacks, because the side pockets and the zipper are exactly like the bermuda shorts on pages 128–133.

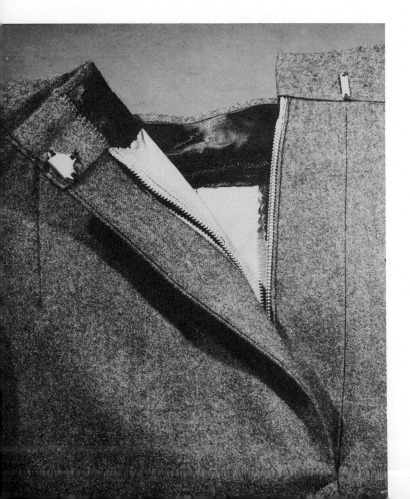

Then, everything else, including the waistband, is done like men's slacks in Chapter 12. The band will vary only in that it finishes at the side instead of center front of slacks. The back pocket is usually eliminated in women's slacks.

These women's slacks were not cuffed, because they are too tapered.

On both men's and women's slacks, use metal hooks and eyes or large snaps.

Chapter 10

Shirts for men, women, and children

Try these techniques on one shirt, and you will have learned them for every shirt. Each one you make will be easier and will go faster. Cut out several and complete the same unit on each one to save time. Cut out a dozen to make for Christmas gifts. A shirt you have made personally will delight young and old alike, and will truly be his or her favorite.

The yoke of the shirt is cut lengthwise. The cuffs and collar are cut lengthwise and on a fold. Firm (drip-dry) siri has been used in the collar, cuffs, and front of the shirt in the photographs. The shirt was made of a medium-weight, textured cotton fabric. In a lighter weight fabric, use a lighter weight interfacing. Shrink siri for garments that will be laundered in hot water.

Buy a man's shirt pattern according to chest measurement. If the man's neck is much larger than the pattern size, increase neck size by cutting neckline larger at the shoulders, and increase collar size.

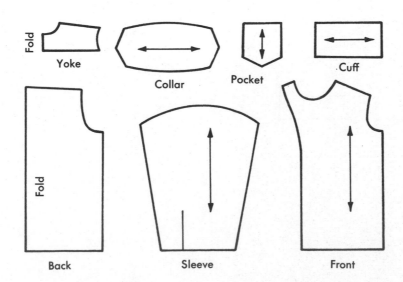

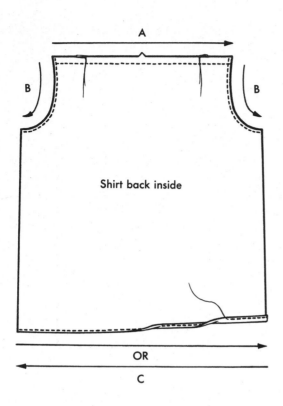

Shirt back unit

The pleats give a better line to the shirt back, when they are placed near the armhole edge.

Fold pleats toward armholes at top of shirt back (A). They are held in place with staystitching, placed just outside the seamline. Staystitch armholes (B) and bottom of shirt (C) ⅛ inch from edge in direction shown. To hem shirt back, turn in raw edge on staystitching line, fold up narrow hem, and stitch at machine along upper edge.

If you make a heavier wool shirt, it is desirable to cut the yoke facing from a rayon twill fabric. The yoke and yoke facing are always cut lengthwise. Staystitch each of them separately at the neckline and shoulders just outside seamline in direction shown.

With right sides together, *and with shirt side up*, sew yoke to shirt back.

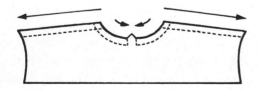

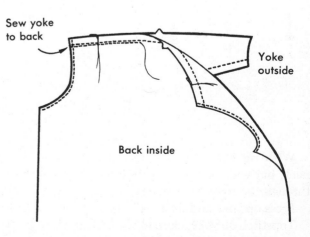

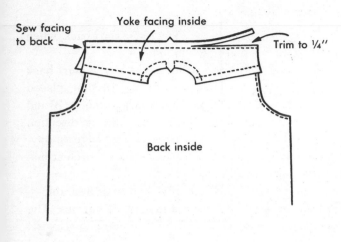

Sew facing to back

Yoke facing inside

Trim to ¼"

Back inside

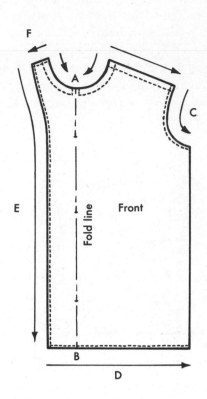

F

A

C

E

Front

Fold line

B

D

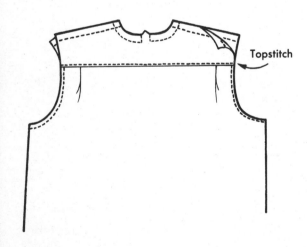

Topstitch

With right side of yoke facing to inside of shirt back, but with yoke side up, sew facing to shirt back through the row of stitching on the seamline.

Press up yoke and facing and trim seam to ¼ inch.

Topstitch on yoke, close to seam, through all thicknesses. In plain fabric, make a second row of topstitching ¼ inch from first row.

Shirt front unit

Clip fold line for facing on shirt front for ⅛ inch at A and B. To press, fold back facing in place; on underside, mark one layer of fabric only with tracing paper and wheel. Through both layers of fabric, place 3 or 4 pins in marking line. Turn back and press top layer on pin line (see photographs, page 164). Remove pins. Turn fabric to have under layer on top. Fold back top layer to meet under fold and press second side. Staystitch neckline and shoulders just outside seamline in direction shown. Staystitch armholes (C), bottom of shirt (D), and edges of facing (E and F) ⅛ inch from edge, and in direction shown. Clean-finish edges E and F of facing. To clean-finish, turn under on staystitching line and stitch close to folded edge (page 74, *The Bishop Method of Clothing Construction**).

*Hereafter referred to as *BMCC*.

137

Tear a lengthwise strip of interfacing (1½ to 2 inches wide) to reinforce shirt front for buttons and buttonholes. It will extend to the top of the shirt and one inch below lower button and buttonhole. Place ⅛-inch from fold line. Stitch in place at edges C and D. At edges A and B, stitching will extend only to top button and buttonhole (E). If edge B is cut on the selvage, it does not require stitching to facing.

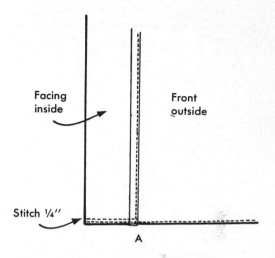

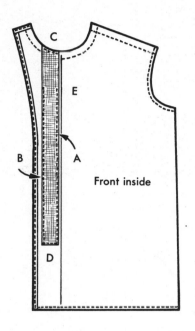

At lower edge, turn facing to outside on fold line. With right sides together, stitch a ¼-inch seam to edge of facing. Trim away corner, turn, and press. On shirt front, at edge of facing (A), clip to staystitching ⅛ inch from lower edge.

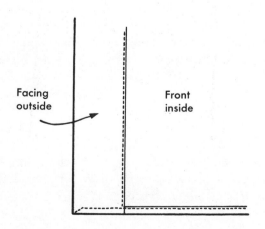

Stitch hem same as shirt back, and continue to topstitch lower edge of facing as shown.

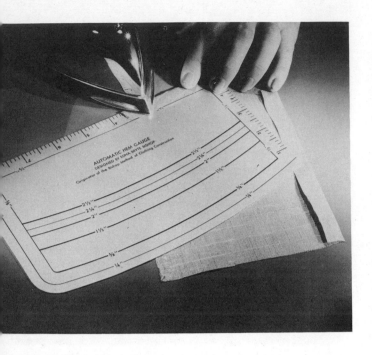

Pocket unit

When there is only one pocket, place it on the left side of the shirt. Never have a rounded lower edge in the style of the pocket when shirt is made of stripes, plaids, or checks. Turn under seam allowance on three sides of pocket and press.

Turn back and press hem allowance at top. Staystitch raw edge ¼ inch from edge; turn in raw edge on staystitching and press. Stitch hem to pocket along lower edge or slipstitch by hand.

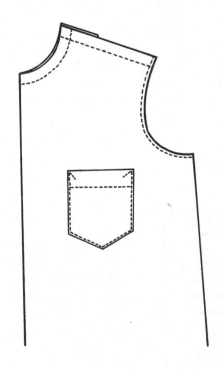

Apply pocket to shirt by stitching close to edge at machine. For reinforcement, at upper corners, stitch back at an angle for ½ inch.

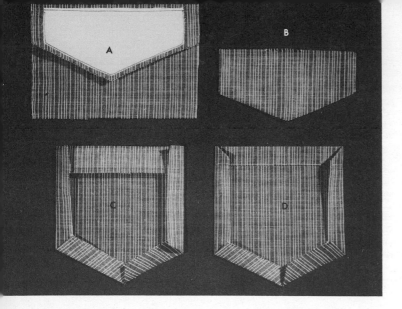

A flap may be added over the pocket.

At A, flap is cut with a fold at the top. Interfacing is stitched inside seamline. With right sides together, stitch around entire flap. Press open one seam edge, and trim double fabric to ¼-inch seam, rounding off seam allowance in corners.

At B, slash fold line for approximately 1½ inches to turn flap. Press and edge-stitch sides and lower edge.

At C, turn in and press seam allowance at sides and lower edge of pocket, and hem allowance at top.

At D, turn under raw edge and stitch hem.

Topstitch pocket to shirt. Reinforce the corners like sketch on page 139. Place right side of flap to right side of shirt above pocket. Stitch along edge and a second row back ¼ inch for reinforcement. Turn down to press in place.

Complete shirt unit

At shoulders, place outside of yoke facing to inside of shirt front. Stitch together on seamline in direction shown (A).

Turn to inside of yoke facing, and place shoulder edge of yoke to shoulder edge of shirt front, right sides together. Do not pin it, but place edges together gradually as they are stitched. Keep yoke facing side up, and sew through stitching on seamline. It is the same technique as the sleeveless blouse on page 36, *BMCC*. Trim seam to ¼ inch, turn, and press.

Topstitch on yoke close to seam through all thicknesses (B). In plain fabric, make a second row of topstitching ¼ inch from first row.

At armhole, trim seam to point at edges of yoke to eliminate bulk in flat-felled seam (C and C). Staystitch armhole edges together ⅛ inch from edge in direction shown (D).

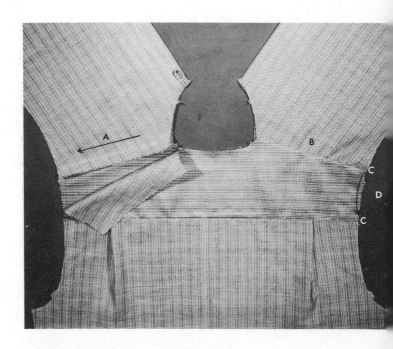

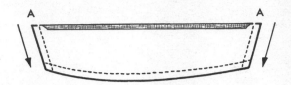

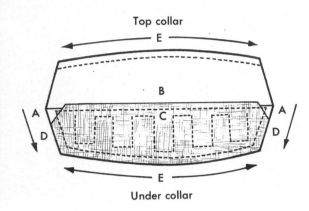

Top collar

E

B

A   C   A

D   E   D

Under collar

Place collar right sides together, and stitch ends of collar in direction shown. Trim interfacing to seamline, and cut open fold of collar to seamline at A and A. Press open seam on edge presser; trim away corner, and seam to ¼ inch. Turn and press collar.

Collar unit

Two other types of collars are shown on pages 150–152 at the end of this chapter.

The collar is always cut lengthwise and on a fold at the outside edge. Press collar in half, right side out.

Cut interfacing identical grain of collar one-half the size of the collar, but wide enough to extend ¼ inch beyond fold line. At corners A and A, cut away interfacing diagonally ¼ inch beyond seamline to eliminate bulk.

Place interfacing toward top collar (B) ¼ inch from fold, but stitch on under collar (C) ⅛ inch from fold. This technique gives support to the folded edge of the collar. Staystitch ends of collar (D and D) on seamline in direction shown. Staystitch neck edges (E and E) just outside seamline in direction shown.

Stitch a design on grain of under collar as shown. This stitching will hold together interfacing and under collar, and will enable collar to launder better and to roll more favorably.

Make a strip of bias tubing to finish ⅛ inch wide (see page 193, *BMCC*); stitch on left edge of shirt front for button loop.

Clip neck edge of shirt at intervals almost to stay-stitching to give a straighter line for stitching (A). Key and pin center back of under collar to center back of shirt, and ends of collar to center fronts of shirt. With shirt side up, stitch under collar to shirt from notch to notch, approximately one inch from ends of collar (B).

Key raw edge of top collar to raw edge of under collar, and sew to shirt just outside seamline from edge of collar (A) to yoke seam (B).

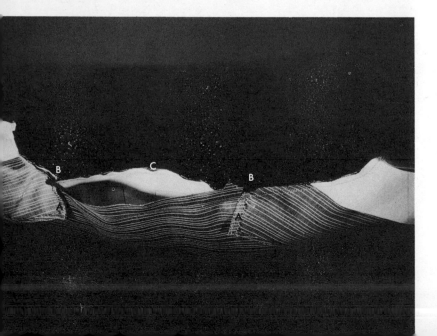

Clip facings at intervals almost to staystitching to give a straighter line for stitching. Place facings on top collar, right sides together; with shirt side up, stitch through all thicknesses to top of facings (A) on stitching on seamline. Trim all of the collar seams to ¼ inch, and clip curves where necessary. Clip into under collar seam at edge of facings (B). Across back of shirt, turn under collar seam (C) up into collar and press.

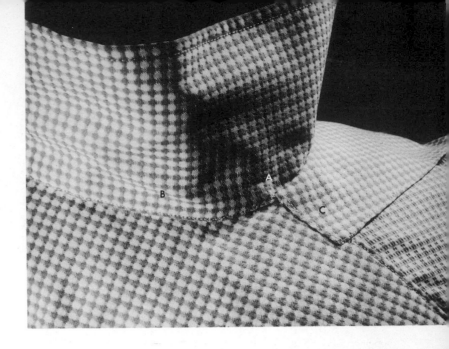

Clip to seamline on top collar at ends of front facings (A). Turn under top collar on seamline and topstitch to seamline of under collar (B).

Stitch edges of facings to shirt with a hand stitch (C) or from outside of garment, stitch facing through well of yoke seam. Press collar.

The shirt collar technique has been simplified here for first learnings on shirts. For a regulation tailored shirt, use the collar technique on pages 104–105, *BMCC*.

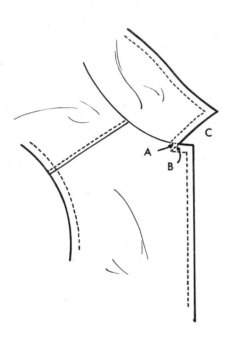

Topstitch fronts and collar of shirt ¼ to ⅜ inch from edge. Some fabrics, such as gabardine and flannel, are hand-picked.

Topstitching is continuous at corners of collar and facings. Topstitching at A sinks into well of seamline. At B, it is done with same angle as edge of collar (C).

Cuff and sleeve unit

The placket opening in the sleeve is made exactly as in the photograph and directions at the top left of page 108, *BMCC*. It was used in the shirt in the photograph at the beginning of the chapter. However, in a fine, dressy broadcloth shirt, for example, the following separate facing technique is used.

Slash opening on true grain of fabric. On narrow side of sleeve, sew on a straight strip of fabric, 1½ inches wide and ¾ inch longer than opening. Have right side of strip to right side of sleeve; stitch ¼ inch from edge and ¼ inch above slash.

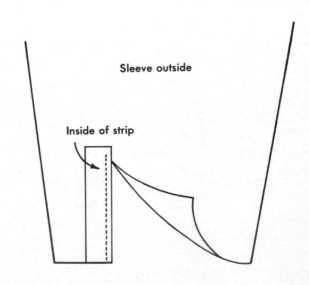

Sleeve outside

Inside of strip

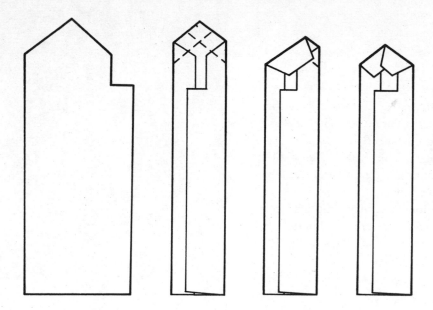

Turn under seam allowance on sleeve facing piece and press. Also, fold and press facing piece on pattern marking line.

Turn in and press ¼ inch on long raw edge of strip. Turn the strip over the slash, lap over first stitching line, and topstitch strip in place on underside. Strip will finish ½ inch wide. Bring sleeve facing piece to right side of sleeve and press in place.

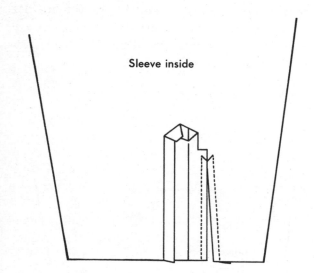

Sleeve inside

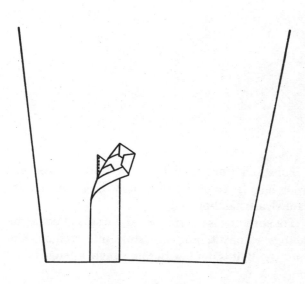

On wide side of sleeve, sew on sleeve facing piece. Have right side of facing piece to inside of sleeve. Stitch ¼ inch from edge and ¼ inch above slash.

Clip to stitching at top of slash, forming a triangle. Press seam on strip toward strip, and press open the seam on the facing piece.

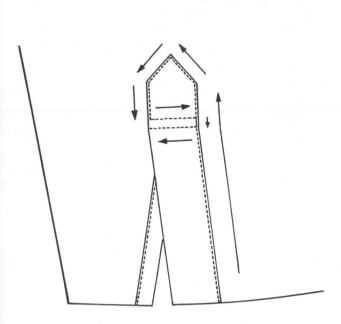

Beginning at sleeve edge, topstitch the side edge of facing piece, point, and back and forth across the top of the opening to conceal and strengthen it.

This photograph shows the finished sleeve opening.

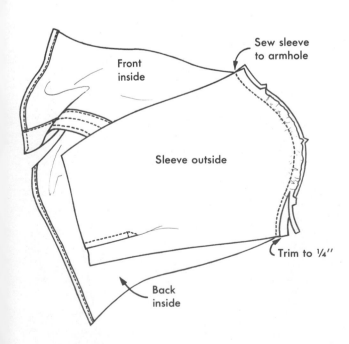

Place inside of sleeve to inside of shirt, and with the sleeve side up, sew sleeve to armhole, easing in sleeve between notches. It is easier to stitch with the outside curve (sleeve) turned up.

Trim only seam of sleeve to ¼ inch. Turn under the shirt seam, clipping staystitching where necessary in curves.

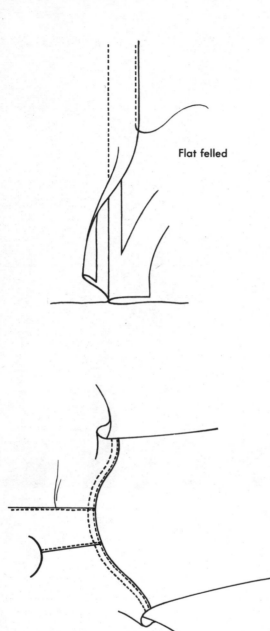

Flat felled

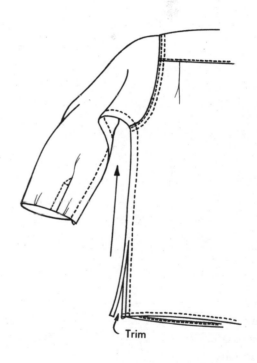

Trim

shirt up. Have shirt wrong side out so that you can sew into sleeves, just as you work up your hosiery on your hands before putting them on.

For the man who is small in the waist, some shaping may be desirable at the sides of the shirt, and the shirt could be tapered in cutting before the side seams are stitched.

Pull fabric apart with both hands and both wrists to be sure the seam is smooth underneath, and stitch shirt seam to sleeve to make flat-felled seam.

With wrong sides together, stitch side seams and seams of sleeve from the bottom of shirt up. Trim only back seam of shirt and sleeve to ¼ inch. Turn under the front seam and sew flat-felled seams on both sides of shirt and both sleeves from bottom of

For the man who is large in the waist, it may be desirable to gradually cut on extra at the side seams at the waistline.

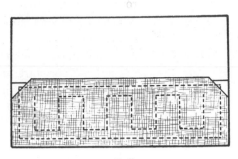

Cuff

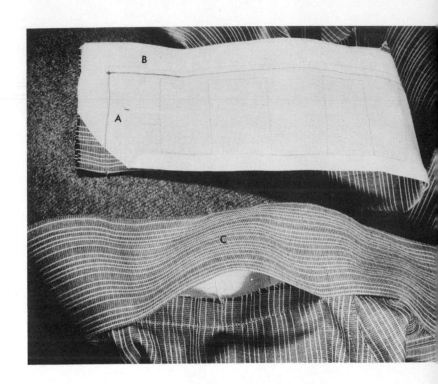

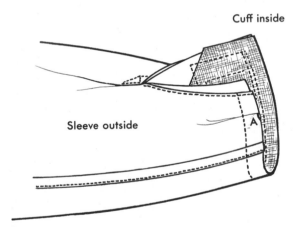

Cuff inside

Sleeve outside

A

The cuffs are always cut lengthwise and on a fold at the outside edge (unless a curved style, etc.). Press cuff in half right side out. Cut interfacing identical grain of cuff one-half the size of the cuff, but wide enough to extend ¼ inch beyond fold line. The cuffs are made precisely like the collar (pages 141–143), including the stitching design with interfacing and under cuff, except that the ends are not stitched together.

To put on the cuffs, place right side of under cuff to wrong side of sleeve. The edges of the sleeve opening will come to seamline at ends of cuff. Any excess sleeve is laid into pleats that are turned toward sleeve and placed near opening (A). Sew with sleeve side up.

Turn ends of cuff right sides together and stitch ends of cuff through staystitching on seamline (A). Continue to sew across top of cuff for one inch through under cuff stitching on seamline (B). Trim interfacing to seamline and seam to ¼ inch; trim away all of seam across corner.

Turn cuff right side out and press (C).

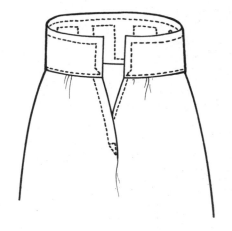

Topstitch close to edge across top of cuff. At corners, stitch back at an angle as shown so that sides and lower edge of cuff can be topstitched ¼ or ⅜ inch from edge, like collar and shirt fronts. If shirt is hand-picked, hand-pick cuffs, also. Finish cuff with a button and buttonhole. See photograph at beginning of the chapter.

Complete shirt unit

Normally, buttonholes are made lengthwise for men and boys, and crosswise for women. If there is a box plait on a woman's shirt, then the buttonholes are made lengthwise. Women's shirts button right over left, but men's shirts button left over right.

Do not make a buttonhole near the neckline of sports shirts for men and boys. The bias loop and small button used with it are all that are needed to hold collar in place at the neckline.

Sew on buttons (pages 84–85, *BMCC*).

Final-press shirt (Chapter 9, *BMCC*).

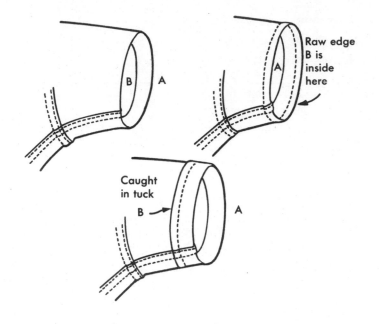

Raw edge B is inside here

Caught in tuck

B ← A

Short sleeve finish

Turn under and press a hem allowance of 1¼ or 1½ inches (A to B). Turn up hem a second time and press on raw edge of first turn.

On outside of shirt, stitch a ¼- or ⅜-inch tuck at top of hem, catching raw edge in stitching. Press tuck up from lower edge.

This is frequently done to shirts when children have outgrown the long sleeves.

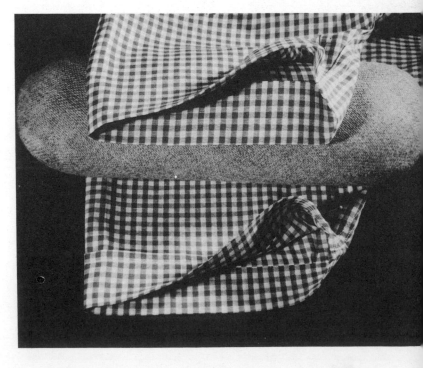

Collar with neckband cut in one

Place interfacing on inside of top collar, and stay-stitch all edges together ⅝ inch from edge, except A, which is just outside seamline.

Make a row of stitching on fold line of collar (B), and three rows below fold line, ¼ inch apart, beginning and ending at the notches (C). These four rows will be continuous, as sketch illustrates.

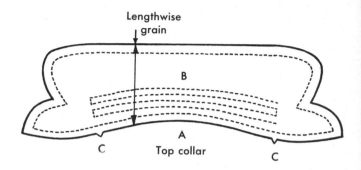

Staystitch edge A of under collar just outside seamline.

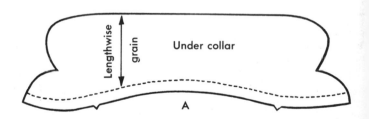

With top collar side turned up, sew on under collar. Trim interfacing to seamline, and seam to scant ¼ inch. Clip into machine stitching at B and C. Clip curves and understitch under collar from D to E, unless entire collar (depending upon fabric) will take topstitching. Turn and press collar.

Stitch through top and under collar on fold line (F).

Collar is put on shirt same as one with separate neckband to follow.

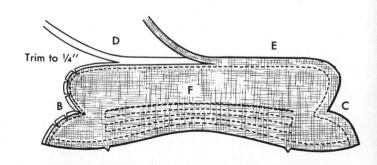

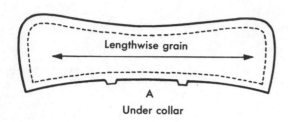

Lengthwise grain

A

Under collar

Collar with separate neckband

Staystitch interfacing to inside of under collar on seamline, except edge A, which is staystitched just outside seamline.

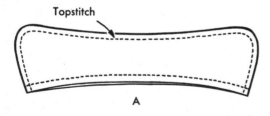

Top collar outside

Under collar inside

With right sides together and under collar turned up, sew on top collar. Stitch through staystitching on seamline. Trim away interfacing to seamline, and seam to scant ¼ inch. Clip curves, turn, and press.

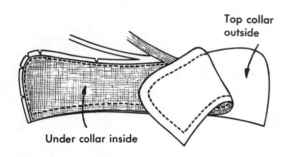

Topstitch

A

On under collar, trim interfacing away to stay-stitching at A, and stitch together raw edges of top and under collar just outside seamline at A. Top-stitch or hand-pick collar.

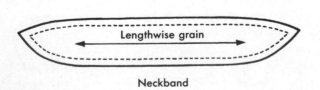

Lengthwise grain

Neckband

Staystitch interfacing to inside of top neckband just outside seamline.

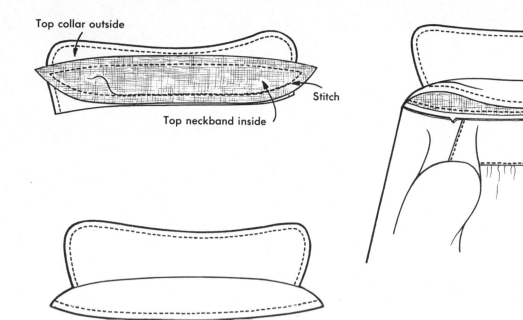

Place top neckband to top collar, right sides together, and with outside curve on top (neckband), stitch band to collar.

Staystitch neck edge of under neckband just outside seamline. Place under neckband to under collar, right sides together, and with top neckband turned up, stitch through same stitching on seamline. Trim interfacing to seamline, and seam to scant ¼ inch, Turn and press collar.

The front facing is pressed in place on the inside of the shirt front. Staystitch facing and shirt together at the neckline just outside the seamline. At shoulder edge, sew facing to shoulder seam of shirt. (If the shirt has a yoke, shoulder edge of facing can be included in the yoke seam, as in bottom photo, page 105, *BMCC*.)

Place right side of top neckband to inside of shirt. With shirt side up, sew top neckband to shirt with a ⅝-inch seam.

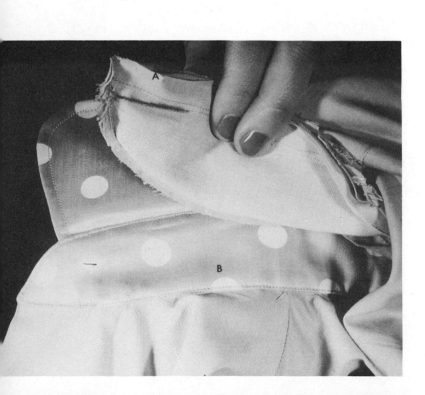

Turn and place raw edge of under neckband to meet raw edge of shirt, right sides together. With top neckband turned up, stitch through stitching on seamline for one inch from edge of collar (A). Repeat on second shirt front.

Trim interfacing to seamline and all of seam to scant ¼ inch. Turn neckband right side out. Press seam up toward neckband. Turn in seam allowance on remainder of under neckband. Edge-stitch opening at machine and continue all around neckband (B).

Chapter 11

Vests for men and boys

A. Red Tweed Vest

C. Scotch-Plaid Tweed Vest in Multi-colored Stripes

B. Striped Flannel Vest

D. Necktie Silk Vest

What husband, what son, what man wouldn't enjoy a classic, well-tailored vest made by you? Choose tweed, plaid, wool flannel (a bright color such as red with brass buttons is always a favorite), or linen for daytime wear or spectator sportswear. Wool flannel is preferred over corduroy, because corduroy does not tailor as satisfactorily. Choose a dressier fabric such as tie silk for evening wear.

A luxury lining such as suracel, rayon twill, satin, peau de soie, or taffeta is desirable.

The entire front is interfaced with firm siri, hair canvas, or Formite, depending upon chosen vest fabric.

To buy the pattern in the correct size, choose one comparable to the size purchased in ready-to-wear.

You will find as you make this vest by the Bishop method that you are using the same basic principles you learned on other garments. Also, unit construction carries over in making every garment. Let us see how our principles are applied now in making a vest.

Front unit

Interfacing for the vest front is cut precisely like pattern for vest front. With tracing paper and tracing wheel, mark pockets on interfacing only.

Staystitch in correct direction of grain all around vest front just outside seamline, except edges A and B, which are staystitched exactly on seamline.

If hair canvas is used as interfacing in front of vest, then a piece of muslin will have to be cut and applied to edges A and B (pages 152–153, *The Bishop Method of Clothing Construction**). With small stitches, staystitch ½ inch each side of corner (C) on seamline.

With contrasting thread, baste-stitch pocket location and size lines through both layers of fabric. Unless four pockets are preferred, mark lower pocket on right front, and upper and lower pockets on left front.

Make regulation welt pockets (pages 174–176, *BMCC*). It is important to remember welts are always made of a lengthwise piece of fabric, and the back pocket piece must be identical to the grain of garment. The front pocket piece is cut of lining and is crosswise grain. The slot pocket on pages 145–146, *BMCC*, may also be used on vests.

*Hereafter referred to as *BMCC*.

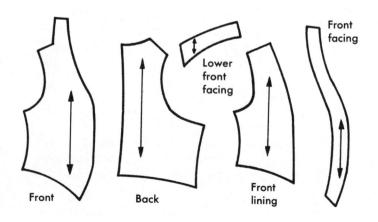

Front   Back   Front lining   Lower front facing   Front facing

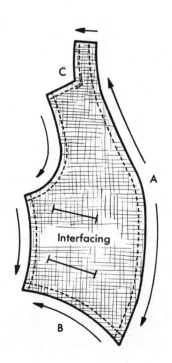

Interfacing

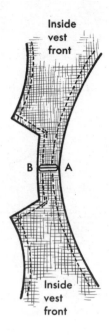

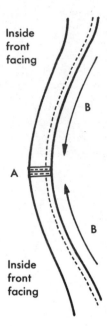

Stitch center back seam of 2 vest fronts. Trim away interfacing to seamline, press open vest seam, and trim to ¼ inch. Beginning 1 inch from edge, trim seam away diagonally to ⅛ inch at A and B to eliminate bulk at seamline.

Stitch center back seam of front facings and press open (A). Topstitch each side of seam, and trim away all of seam up to topstitching.

Facing will be stitched to vest in next step with vest side up. To keep edge grain perfect, staystitch off-grain edge B of facing *at neckline only*, just outside seamline.

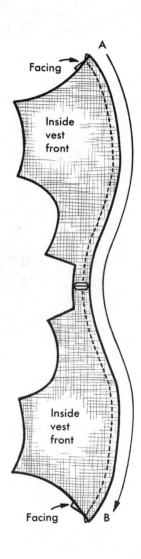

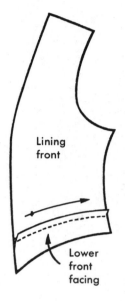

Stitch lower front facing to bottom edge of lining front in direction shown. Press seam up; leave wool seam ⅝ inch and trim lining seam to ⅜ inch.

With right side of facings and vest fronts placed together, and with vest side up, sew facing to vest fronts through staystitching on seamline all the way from one lower front edge (A) to the other (B). Use smallest stitch at machine for a distance of one inch at lower edges.

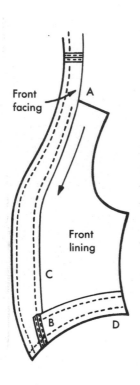

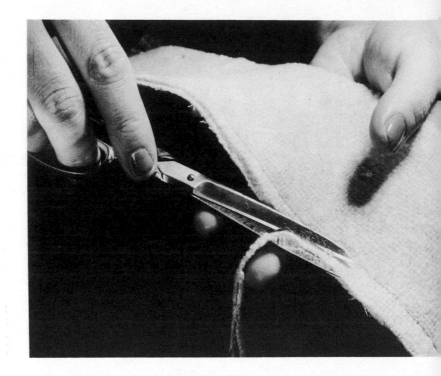

Beginning one inch from top of lining front at A, stitch inside edge of front facing to lining front and lower facing. At B, press open seam, topstitch each side of it, and trim away all of seam up to topstitching. Press remainder of seam (C) toward lining front. Leave wool seam ⅝ inch and trim lining seam to ⅜ inch.

With right sides together and with vest side up, sew lower facings (D) to vest fronts through staystitching on seamline. Use smallest stitch at machine for a distance of one inch at front corner.

At front and bottom edges of vest, trim interfacing to seamline. Then, hold scissors perfectly flat for trimming, and stagger both seam edges of vest at once, making one toward top of vest ¼ inch and under one ⅛ inch. Round off seam allowance at corners to within a few threads of stitching line.

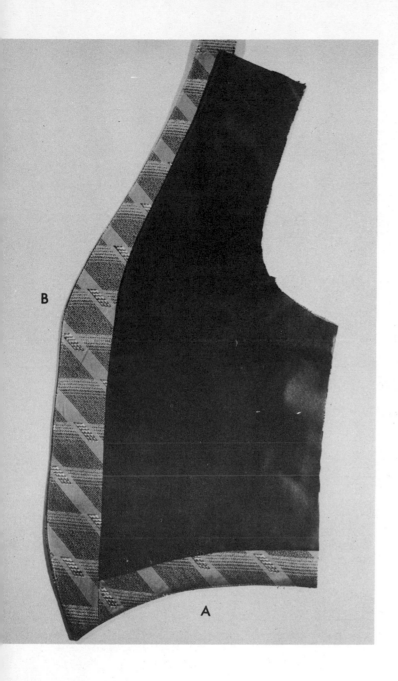

Turn vest right side out. Understitch bottom edge (A), beginning in corners as far as possible to one inch from side edge of vest. Understitch front edges of vest (B), also going into corners as far as possible. Understitching is a row of machine stitching along the edge of the facing that catches the two trimmed seams to the facing. Press edge from top side of vest.

Place right sides together and stitch lining front to vest front at armhole in correct direction of grain. Trim away interfacing to seamline and lining and vest fabric to ¼ inch. At intervals, clip curved armhole seam to seamline. Turn right side out, and beginning as far up armhole as possible, understitch armhole seam to side edge. Press armhole edge from top side of vest.

This photograph shows the completed front unit from the top side.

Back lining unit

Stitch center back seam of upper back lining and under back lining in direction shown and press open seams. Staystitch neckline, shoulders, armholes, and sides just outside seamline in direction shown.

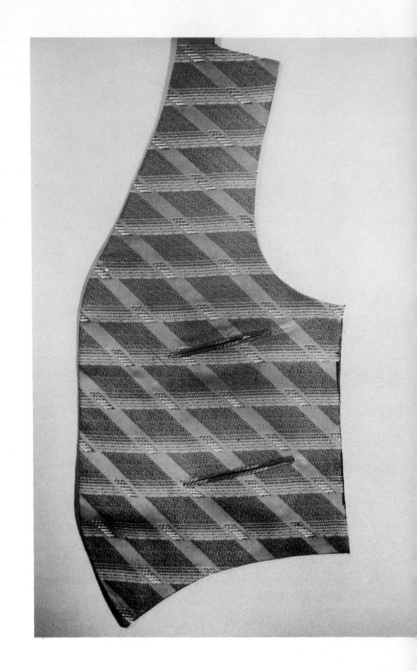

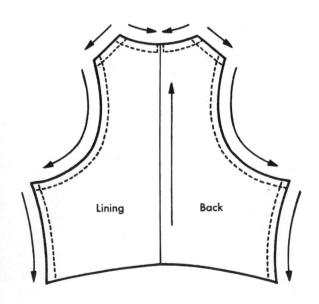

Lining   Back

Then, place lining backs right sides together, and stitch armholes (A and A) and lower edge (B) in direction shown. Trim A and B seams to ¼ inch, and clip curves of armholes. Understitch lower edge, beginning and ending one inch from side edges. Turn right side out and press from topside of lining. Instead of understitching armholes as front ones were done, topstitch them close to edge for added strength, *after* vest is completed.

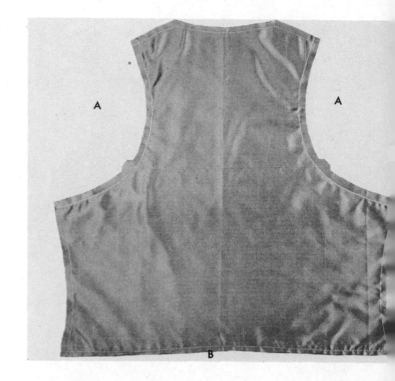

Complete vest unit

Clip corner (A) on shoulder of vest front up to stay-stitching on seamline. Open up shoulders, and pin and stitch right side of lining front to right side of under lining, and on around to right side of vest to right side of upper lining. Continue across neckline, joining vest front and upper lining; do second shoulder seam same as first. Sew with vest side up and use short stitches in corners. Open up armhole seams when stitching over them.

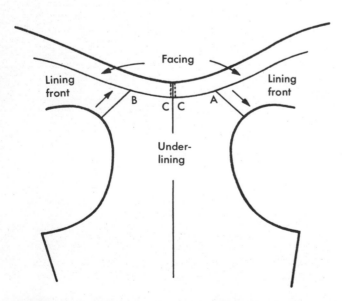

Stitching will begin at A, and continue around to B. The only part that will not be stitched at the machine is neck edge (C) of under lining and facing.

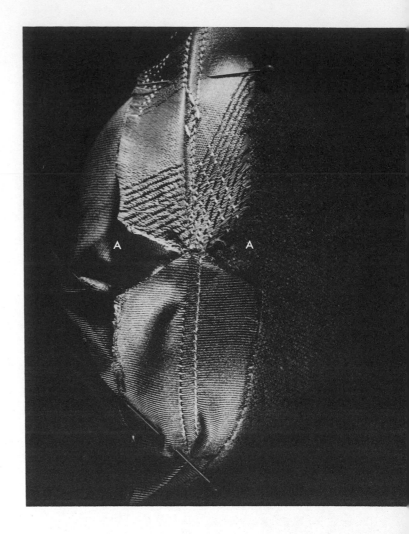

Trim away shoulder seams diagonally as shown at A and A.   Trim all of seam to ¼ inch and press down toward back lining.

Whipstitch back lining to neck facing by hand. (This technique is the same as the blouse, dress, or jumper with neck and armhole facing cut in one (pages 35–36, *BMCC*).

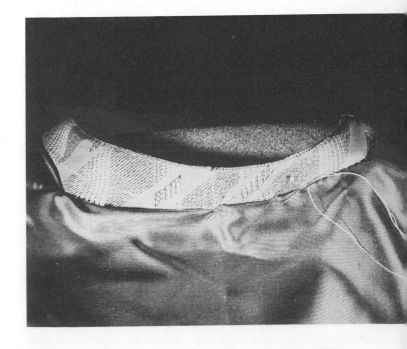

Turn lining wrong side out, and at the side seams, place right side of vest to right side of upper lining, and right side of lining in front vest to right side of under lining. With a continuous line of stitching and a ⅝-inch seam, stitch a complete circle. Turn right side out, after clipping seam to a point at armhole and lower edges. Press seam toward back of vest.

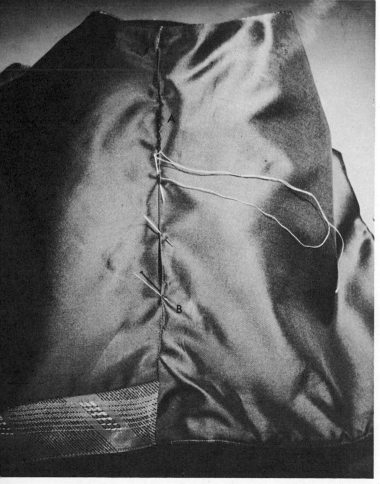

Repeat for second side, except an opening must be left in the middle third of under lining to turn right side out (A to B). Whipstitch same as neck edge. Topstitch back armholes close to edge for added strength.

Make 4 or 5 keyhole buttonholes in left front no longer than ⅝ inch. The vest will not be complete until it has had a professional pressing (Chapter 9, *BMCC*). Sew buttons on right front with a shank (pages 84–85, *BMCC*).

Chapter 12

Slacks for men and boys

Many of the techniques for men's or boys' slacks are identical to those for women's. Even in fitting them with perfection, it is just as important to have the grain evenly balanced on the figure. Slacks do not take much time to make—cut out several pair before you go to the machine to begin to sew. These techniques are used on boys' slacks, beginning with school age. It is assuredly a place where you can save much money in the family budget, and moreover, have slacks with a handsome, quality look.

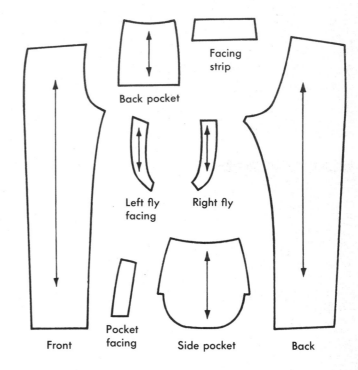

Make them in flannel, sharkskin, whipcord, corduroy, wide whale cotton, wash-and-wear fabrics, or other desirable ones of your family's choice.

Buy the slacks pattern by the waist measurement.

Back unit

Sometimes, the front of the slacks may need to be increased in width for the individual figure and not the back; or the reverse may occur, the back will need to be increased and not the front.

Staystitch top of slacks and sides for a depth of ten inches just outside seamline, and in correct direction shown.

To make creases in slacks, on underside mark crease line on one layer of fabric only, with tracing paper and wheel. Through both layers of fabric, place 4 or 5 pins in marking line.

Turn back and press top layer on pin line. Remove pins. The crease begins 8 inches below waistline on back of slacks.

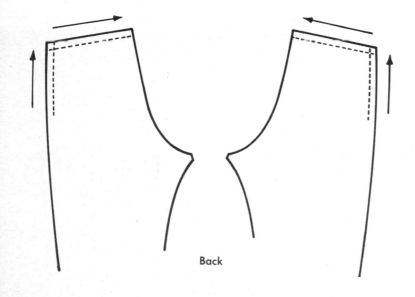

Back

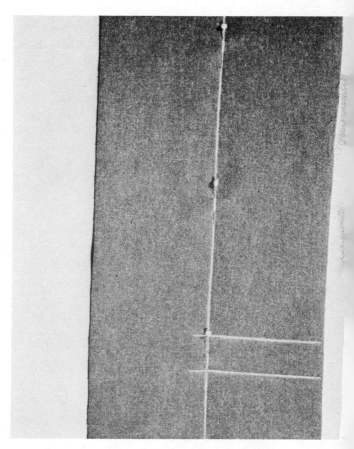

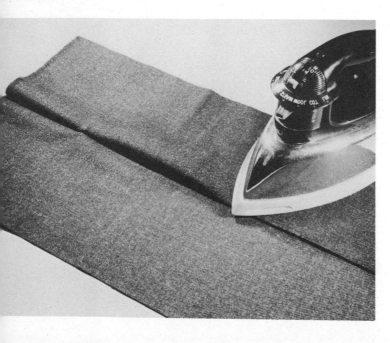

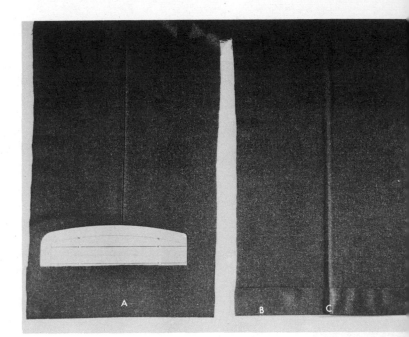

Turn fabric to have under layer on top now. Fold back top layer to meet under crease and press second side.

Turn up hem at lower edge to underside and press (A). Then, turn up cuff on cuff line and press (B). Finally, fold and press crease line again at lower edge to restore it (C).

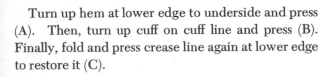

Sew seam at crotch in direction shown. Considerable strain is placed on this seam; strengthen it by stitching with machine set one point or even less to zigzag. If you do not have a zigzag machine, sew with shortest stitch at machine.

Set the machine more than one point to zigzag and stitch both sides of seam ¼ inch from edge. Stretch seam in curve only. If you do not have a zigzag machine, staystitch seam, and stretch in curved area, also. Press open seam.

Stitch darts and press toward center back.

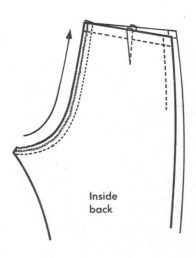

Inside back

Slacks for men and boys

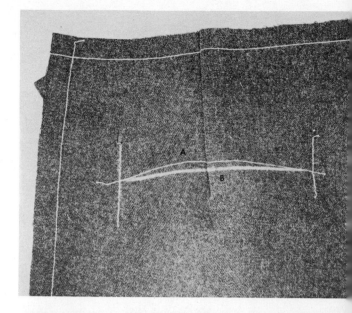

Back pocket

If you just make one pocket on the back, have it on the left side.   You may need to straighten location line for pocket (A) after the dart is stitched.   B is the newly drawn pocket location line.

This pocket uses many of the steps from the modulated welt pocket, pages 116–117, *The Bishop Method of Clothing Construction.* °

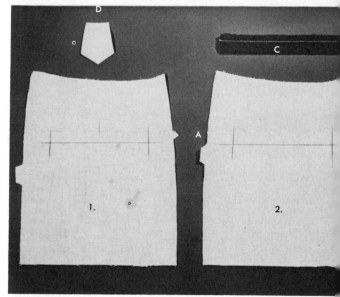

Mark pocket location lines and size lines on muslin pocket pieces (1 and 2).

To make the welt, tear or cut a lengthwise strip of fabric 1½ inches wide and as long as pocket piece is wide (A to B).   Fold welt lengthwise (C), wrong sides together; press; and stitch along the edge, allowing ¼-inch seam.   Welt will finish ½ inch wide.

If the little tab is desired coming out of top of pocket (see photograph on page 168), make it double of slacks fabric, turn, and press.   Leave open D edge to turn right side out.

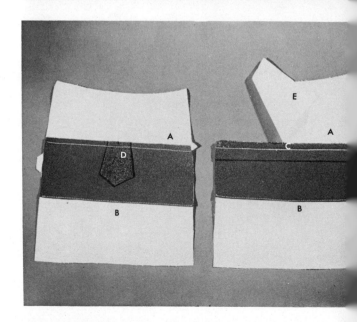

Stitch facing strips to pocket marking lines ¼ inch from raw edge of facing strips (A and A).

Staystitch lower edges of facing strips (B and B) ¼ inch from edge.   Turn under ¼ inch to staystitching and press; topstitch to muslin pocket pieces.

Then, sew on welt (C) through same stitching ¼ inch from edge.   Stitching *must not* extend beyond pocket size lines.   If tab is being used, stitch in place ¼ inch from edge also (D).

Trim away muslin pocket piece above raw edges of welt (E) on this section *only*.

° Hereafter referred to as *BMCC.*

Place raw edges of welt to pocket marking line on right side of slacks, and with welt side down, sew on welt through previous stitching to *size lines only* (A).

Key raw edge of facing strip on second pocket piece to pocket marking line on right side of slacks, and with tab turned down, sew on second pocket piece through previous stitching to size lines only (B). The muslin was rolled up for the photograph to show lower pocket piece, but will be out flat for stitching. Check for accuracy. The stitching lines should be ½ inch apart and must end at the size lines.

Remove baste-stitching at location and size lines in slacks. On inside of slacks, starting at center, slash between stitching lines and diagonally to corners, leaving triangles ½ inch long at ends of opening. Turn pocket sections to inside; press pocket. Welt will fit with perfection in opening.

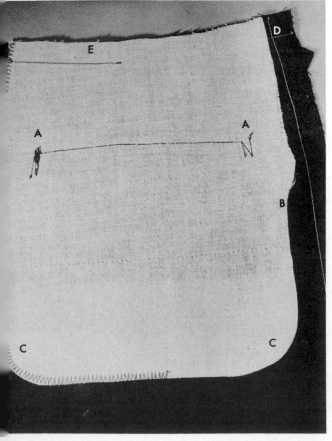

Stitch the triangle to the pocket sections, going back and forth many times to fasten securely. This connecting line at base of triangles (A and A) also squares the ends of the pocket.

Trim any uneven edges that do not fit together with perfection. With wrong sides together, stitch sides and bottom in ¼-inch seam, going up sides as far as possible (B and B). Round off corners in stitching and trim them away (C and C).

Turn pocket pieces and press edges. Zigzag edges of pocket, going all the way to the top of it (D and D). Turn under single raw edge ¼ inch as you zigzag above B and B. Trim away ¼ inch on welt; it must not be turned in at that area. Stitch muslin pocket in place at top of slacks (E).

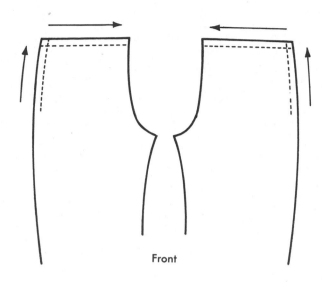

Front

This is the completed pocket. For security reasons, there would be a buttonhole in the tab, and the button would be sewn on the trousers.

Without the tab, the button and buttonhole are optional. Yet, it is preferred to lock in the wallet.

Front unit

Staystitch top of slacks and sides for a depth of ten inches just outside seamline in correct direction shown.

Make creases and cuffs in slacks the same as you learned with back unit on pages 164–165.

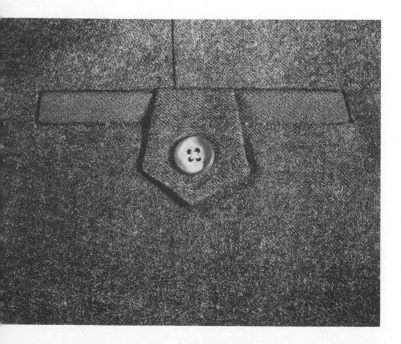

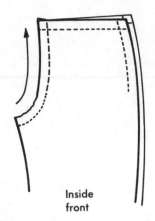

Inside
front

Sew seam at crotch in direction shown. It will be permanent stitching up to the metal on the zipper; lockstitch threads, and then baste-stitch seam for length of metal part of zipper plus all of the tape at the top. Special trousers' zippers are available.

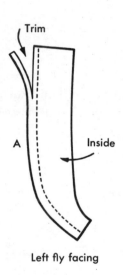

Trim

A   Inside

Left fly facing

On left front fly facing, at edge A, stitch on a piece of lightweight fabric cut same as facing. Trim seam to scant ¼ inch; turn and press.

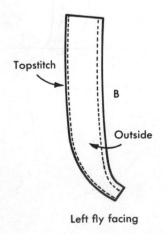

Topstitch

B

Outside

Left fly facing

Topstitch close to edge for added strength. Stitch together raw edges at B.

The entire fly is made in such a way that you just work with the two parts of the center front seam in following these steps.

On left front, place edge B of facing to meet raw edge of left front seam, right sides together.

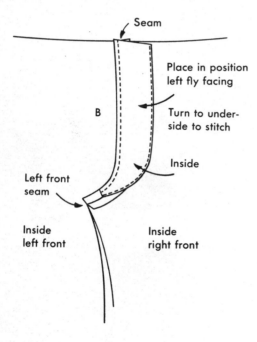

Seam

Place in position
left fly facing

Turn to under-
side to stitch

B

Inside

Left front
seam

Inside
left front

Inside
right front

Turn to underside of left front seam and sew on facing as near basting line as possible.

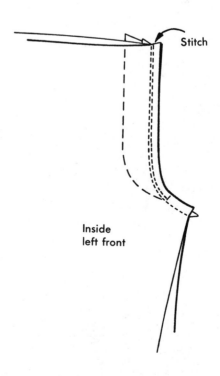

Stitch

Inside
left front

Stagger edges in trimming and understitch edge of
facing. Photograph illustrates steps to this point.

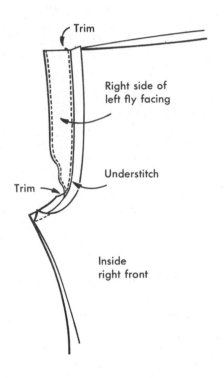

Trim

Right side of
left fly facing

Understitch

Trim

Inside
right front

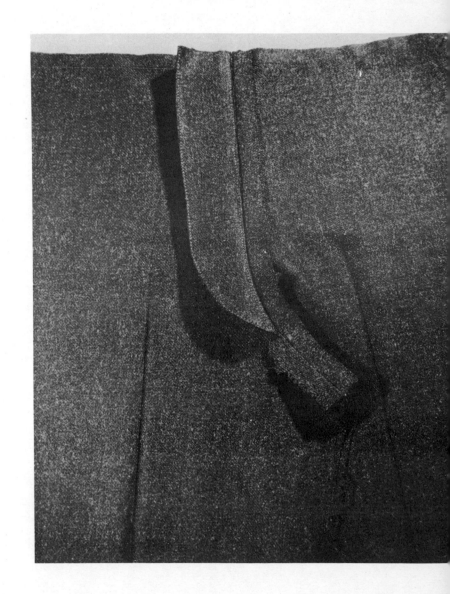

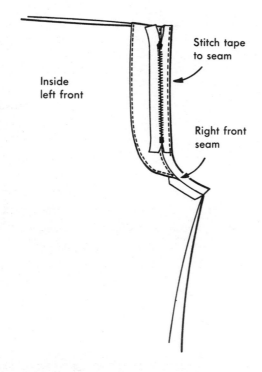

Inside
left front

Stitch tape
to seam

Right front
seam

A

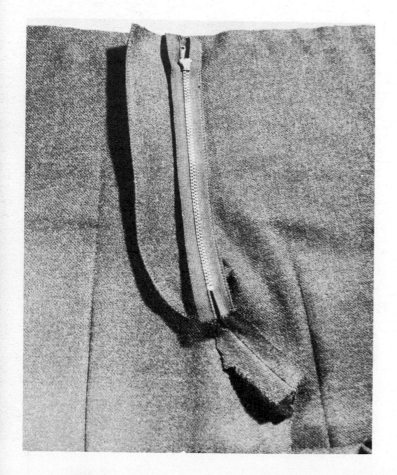

With right side of zipper to right side of right front seam, sew zipper to seam from top down to lower edge. Key edge of zipper tape to edge of seam, and with regular presser foot, stitch close to edge, having edges of tape and seam in middle of right side of presser foot (A).

The photograph illustrates this step.

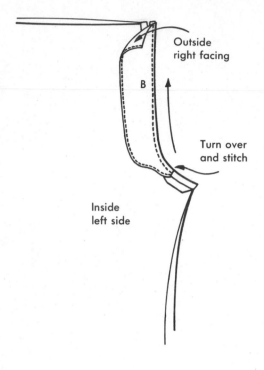

Outside
right facing

B

Turn over
and stitch

Inside
left side

Make right fly exactly as you made left fly facing.

Place right side of facing (edge B) to meet edge of zipper tape. Turn and have underside of right front seam on top, and stitch from bottom up near same line of stitching that sewed on zipper tape.

Turn zipper right side up, roll right front seam up to chain, and with zipper foot, sew close to zipper chain on right front seam. The right facing is out flat.

The photograph shows this step completed.

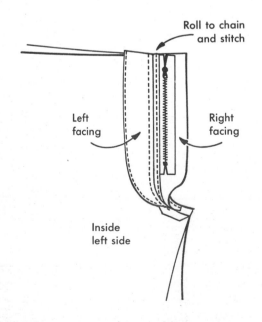

Roll to chain
and stitch

Left
facing

Right
facing

Inside
left side

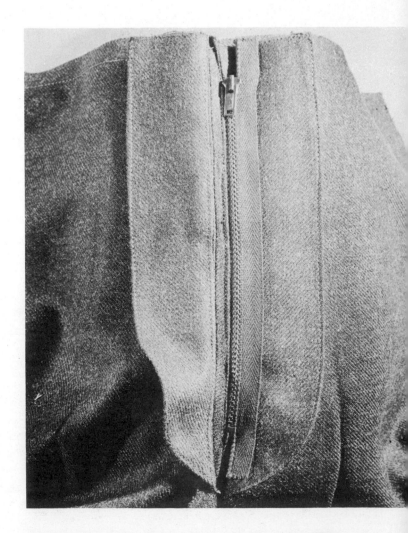

Lay zipper flat toward left front of slacks. Press flat on underside and topside of slacks at seamline. Pin zipper tape to left front fly facing. From bottom up, stitch close to edge of zipper tape with regular presser foot. Put on zipper foot and make a second row of stitching close to metal chain.

Note the photograph of this completed step.

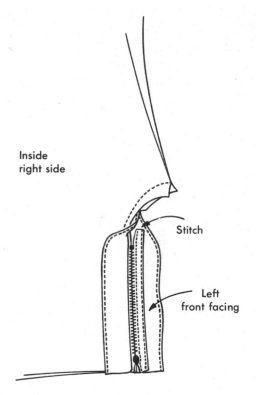

Inside right side

Stitch

Left front facing

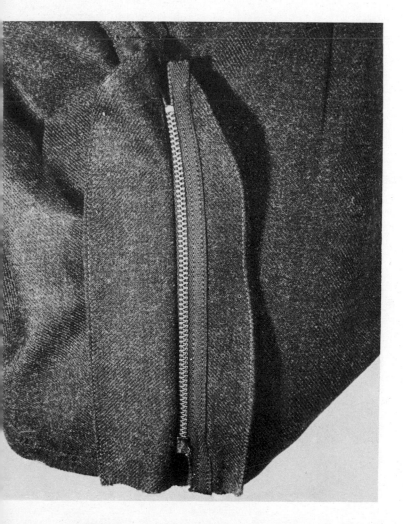

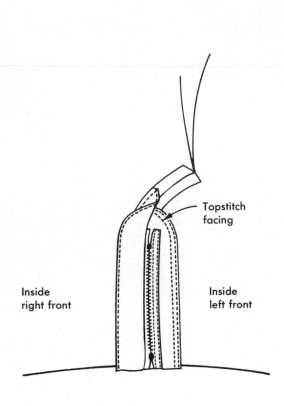

Topstitch
facing

Inside
right front

Inside
left front

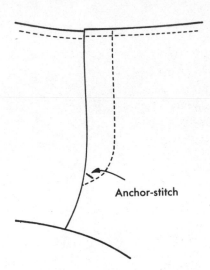

Anchor-stitch

Anchor-stitch at an angle for ⅝ inch at lower edge of opening. Use zigzag stitch at machine. If you do not have a zigzag machine, stitch back and forth several times to anchor.

Turn to underside of right front seam to remove baste-stitching.

On left front, topstitch facing to trousers. Stitch from bottom up ½ inch from edge of facing.

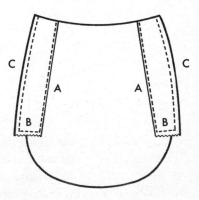

Side pockets unit

On pocket facings, turn in ¼ inch on edge A and press. Pink edge B. Key raw edge of pocket facing at C to raw edge of muslin pocket. Stitch raw edges in place at B and C ¼ inch from edge. Close to the edge, topstitch edge A to muslin pocket.

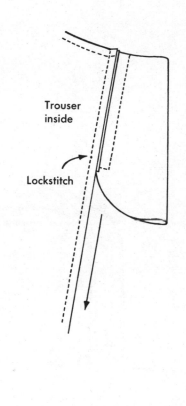

Trouser inside

Lockstitch

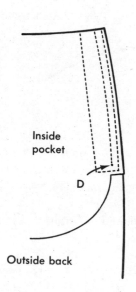

Inside pocket

D

Outside back

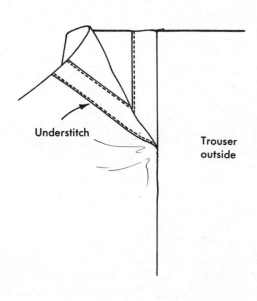

Understitch

Trouser outside

With right side of muslin pocket facing right side of the back of the slacks, sew pocket to slacks with pocket side up. End stitching ⅝ inch from lower edge of pocket (D), and lockstitch threads securely. Repeat for second pocket.

Then, do this same step and sew pockets to front of slacks.

Key together lockstitching on side front and side back of slacks. Beginning at lockstitching, close side seams, stitching down to bottom of slacks.

On front of slacks, trim away muslin close to seamline at pocket edge. Stagger edges of slacks' seam and understitch (see photograph of vest, page 157).

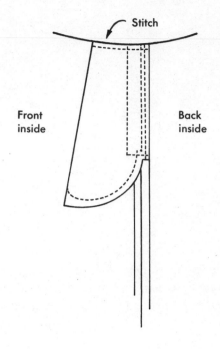

Stitch

Front
inside

Back
inside

Match upper edges with upper edge of slacks, and
stitch together.

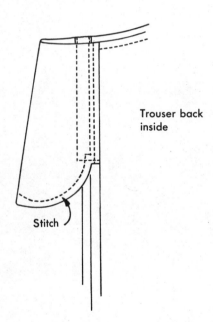

Trouser back
inside

Stitch

On back of slacks, press pocket seam toward back
of trousers.

Below pocket, press open seam.

Key together lockstitching at lower edge of pockets,
and beginning exactly at lockstitching, close pocket
pieces to waist edges.   Round off any corners to keep
out lint.

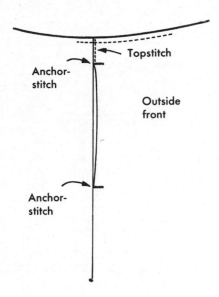

Anchor-stitch straight across for ½ inch at lower edge of pocket opening and 1⅝ inches from top. Top-stitch pocket close to edge from upper anchor to top of slacks.

This photograph shows a finished pocket on the completed slacks.

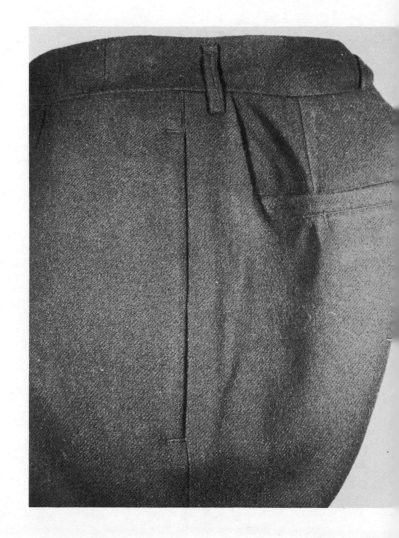

Band unit

Always make straps from a lengthwise strip of fabric ¾ inch wide, having one edge on the selvage. Press back raw edge ¼ inch and press selvage over it ¼ inch. This makes straps ¼ inch wide. Stitch at machine through center of strap.

Place straps around top of slacks where desired, and stitch to raw edge (A). The underside of the strap is placed to the right side of the slacks.

The band to finish 1½ inches wide is cut of a lengthwise strip 2¾ inches wide and 3½ inches longer than waist measurement.

Interfacing of muslin is also lengthwise and as long as the band, but is only 2 inches wide.

Place interfacing ¼ inch from back raw edge of band, and stitch together near edge of interfacing. Press band, rolling in place from back edge toward front edge.

Roll over ⅝ inch to extend on back of band. Stitch near edge of interfacing wherever it falls on front edge of band.

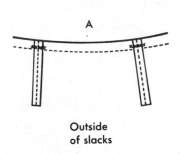

A

Outside
of slacks

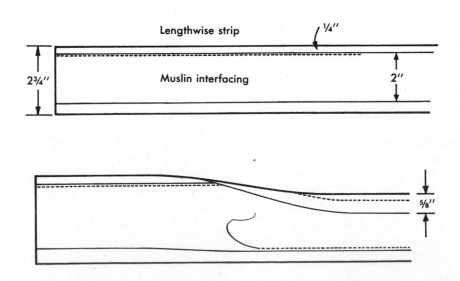

Lengthwise strip   ¼"

2¾"   Muslin interfacing   2"

⅝"

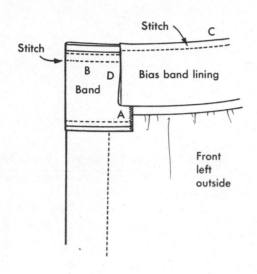

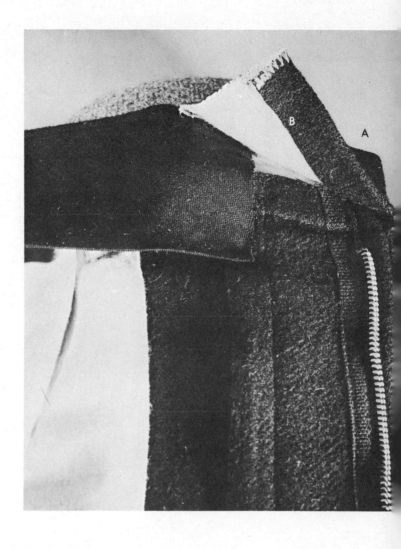

For tolerance needed in slacks, pin band to slacks with the same techniques you learned on the skirt (page 43, *BMCC*). The band extends a seam allowance at right front fly facing, and several inches extra at left front fly facing.

With right sides together, and with band side up, stitch band to top of slacks, catching straps in stitching.

Press open band seam at waistline and trim muslin to seamline from pockets.

At left fly facing, zigzag A edge first. Turn band extension to right side of band and stitch on seamline (B).

Cut bias muslin (or rayon twill in matching color of slacks) 5 inches wide. Press in half and stitch together raw edges ⅜ inch from edge. Place raw edges to meet raw edge of back of band (C). At left front fly facing, lap muslin ½ inch over band extension (D). Beginning at raw edge of band extension and with band on *top*, sew muslin to band ⅜ inch from edge all the way across to raw edge of band at right front fly facing. Turn to inside and press seam down. Muslin will be ¼ inch from top of band.

On band extension, press open top seam on edge presser (A). Stagger width of seam in trimming and round off seam allowance in corner. Turn right side out and press.

Turn under lower edges of band extension and whipstitch by hand to seamline. Raw edges may be staggered first in trimming. At end of extension, clip lower waistline seam up to stitching.

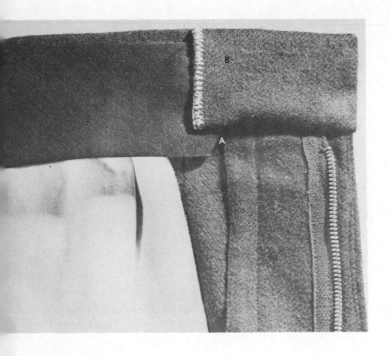

Continue topstitching on left fly facing to top of band.

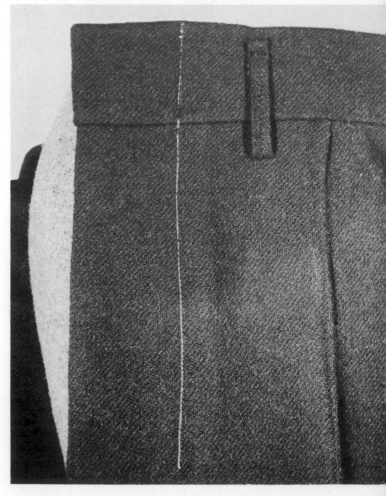

Turn under raw edge of bias (A) and stitch in place. B edge will be held in place with topstitching, as in next photograph.

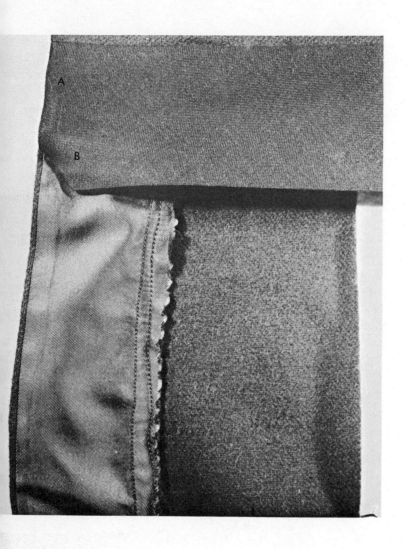

At right front fly facing, turn bias muslin to outside on fold line at top of band. Beginning at edge of facing, sew ends together parallel to edge of facing (A). Use small stitches in the corner for ½ inch. Stagger width of seam in trimming, and round off seam allowance in corner. Turn right side out and press.

Trim away lower part of waistline seam at an angle. Turn in bias muslin diagonally to cover raw edges (B).

Topstitch this edge as shown.

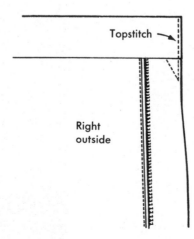

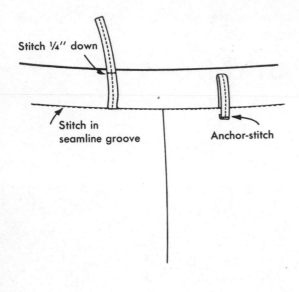

Stitch ¼″ down

Stitch in seamline groove

Anchor-stitch

With right side of slacks turned up, stitch bias in place at lower edge with a concealed row of stitching in the seamline groove at waistline.

The fold of the bias extends to cover up the raw edge of the waistline seam. Final-press band.

Stitch loops ¼ inch from top edge of band. Fold over and bring ½ inch below waistline seam. Turn under raw edge ¼ inch and anchor-stitch in place. This technique makes two sizes of loops, one for a narrow belt and the other for a wide belt.

Sew on hook and eye (see photograph, page 134).

Cuff unit

Staystitch raw edge ¼ inch from edge and pink with pinking shears.

Stitch two inside leg seams from the crotch down, and press open. Trim all seams in hem of cuff to ¼ inch, going ⅛ inch beyond fold line of hem (see C in top photograph, page 162, *BMCC*).

Press hem line and cuff line again to restore creases.

Stitch hem in place as shown at the bottom of page 46, *BMCC*.

Hold cuff in place with concealed hand stitching at two seams.

A firm rayon may be stitched into the seams at the knees of slacks to help retain their shape and reduce wear.

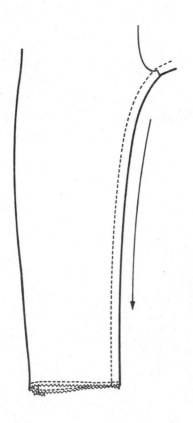

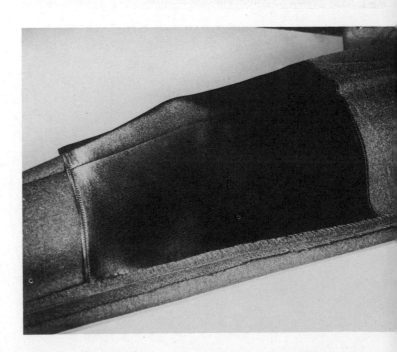

Chapter 13

Sports jackets for men and boys

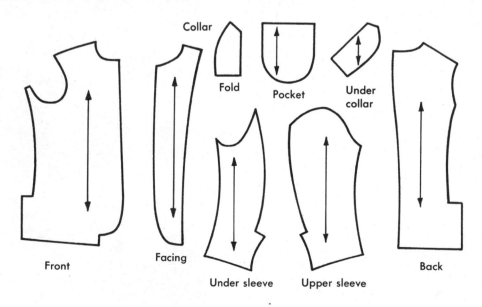

Collar

Fold

Pocket

Under
collar

Front

Facing

Under sleeve

Upper sleeve

Back

Making sports jackets for various members of your family will bring you much pleasure and satisfaction. The techniques presented in this chapter will enable you to construct a jacket with high standards and a fine, quality appearance. The learnings will be identical for making men's or boys' topcoats or overcoats. To buy the jacket pattern in the correct size, choose one comparable to the size purchased in ready-to-wear.

Cutting interfacing

After cutting and marking the fabric for the jacket, remove pattern pieces, and place the sections of the jacket in units of work.

Unless the pattern company has given separate pattern pieces for cutting interfacing *exactly* like the directions to follow, use the original pattern pieces for jacket to cut interfacing. It is always cut on identical grain as jacket.

Hair canvas or Formite. Hair canvas was used in the heavier tweed jacket in the photograph, and Formite was used in the lighter weight jacket done in striped flannel.

Interfacing for the front of the jacket is cut two inches wider at the bottom than width of facing (A). Underarm, it is cut to first marking line for dart (B). Cut *unbleached muslin* for remainder of underarm interfacing, beginning at second marking line for dart (C).

To prevent the front of the jacket from breaking at the armhole, cut an extra piece of hair canvas (A) that will be stitched and quilted on front interfacing (see page 194). At B edges, cut ⅞ inch smaller than edge of tissue. If needed, several layers may be used, or Pelomite may be pressed on this piece. Stagger any additional layers in size.

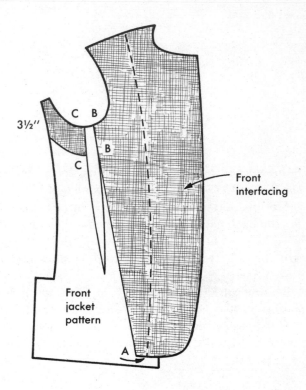

3½"

Front interfacing

Front jacket pattern

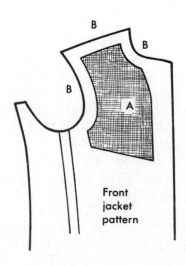

Front jacket pattern

185

The interfacing for the collar is cut from hair canvas or Formite, using the pattern for the under collar. It will be in two pieces and on the bias.

To cut the interfacing for the bottom of the sleeve, see page 200.

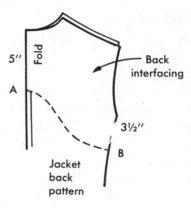

Collar interfacing

Unbleached muslin. To preshrink, see page 11, *The Bishop Method of Clothing Construction.*°

Muslin interfacing for the back of the jacket is cut like the back pattern from A (5 inches down center back) to B (3½ inches below armhole). Remove pattern and cut freehand from A to B, as shown with broken line.

Because you have straight of grain, you may cut the muslin on a fold and omit the seam at the center back.

This is especially advisable for heavy fabric. See photograph on page 190 for lightweight fabric.

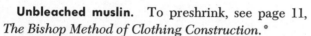

Cut a strip of muslin from front pattern piece on identical grain of jacket front. It is used at the edge where the facing will be attached, and extends to the center front at neckline. This eliminates the hair canvas from the seamline, holds interfacing securely, and makes possible sharp, thin front edges. Place front tissue pattern on identical grain of muslin and cut edges A. Remove pattern. Measure and mark line B, 1¼ inches from front edge A, and cut along B. Unless you have very soft wool, a strip of muslin is not necessary on front edge of Formite interfacing. Staystitch interfacing to garment on seamline at this area.

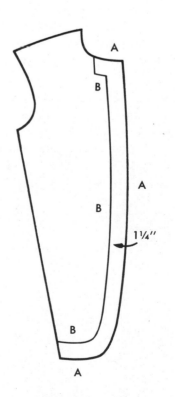

*Hereafter referred to as *BMCC*.

186

Enough true bias of muslin 2½ inches wide is needed to cushion the hem of the jacket. When sewing together strips of it, overlap ⅛ inch on true grain, and stitch at machine.

Cutting lining

It is best to wait to cut the lining until the jacket has been sewn together and tried on for a fitting. If any cutting-to-fit alterations were necessary in cutting out the jacket, cut the lining with the same alterations. Then, if any fitting is necessary when the jacket is tried on, cut and stitch lining with the same alterations.

The original pattern used for cutting the garment is preferred for cutting lining. The cutting and fitting alterations are more easily made, and in carrying out special directions below for cutting, the lining will fit the garment with perfection.

Sleeves are cut to the finished length of jacket sleeve plus ¾ inch extra for ease in lining. Omit sleeve pleat. The ¾ inch extra for ease is not necessary on *boys'* jackets.

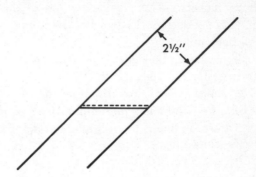

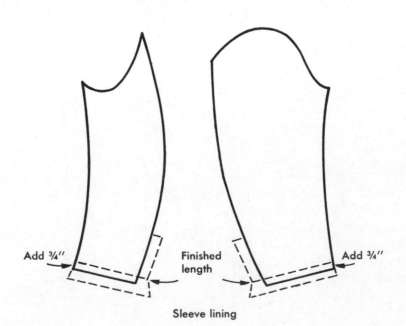

Sleeve lining

The front and back of the lining are cut to the finished length of the jacket plus ¾ inch extra for ease. As above, the ¾ inch extra for ease is not necessary on boys' jackets. The pleat extension at the side opening is needed on the front of the lining, but is cut off back lining.

Chalk mark, on the jacket front tissue pattern, the width front facing will extend on it (line A). Measure 1¼ inches (2 seam widths) from line A toward front of jacket, and chalk mark (line B). Fold under tissue pattern on line B to cut lining for jacket front.

One inch extra width is added at center back of lining down to the waistline for a pleat to give necessary ease. If the back of the jacket is not cut too much off grain, leave edge A on the fold to the waistline, even though the pleat will not be a true one-inch width all the way down.

With the same slope as the shoulder line, cut on 1½ inches of lining at armholes front and back, tapering to ¼ inch extra at notch. Continue to add ¼ inch at armhole from notch to side of jacket.

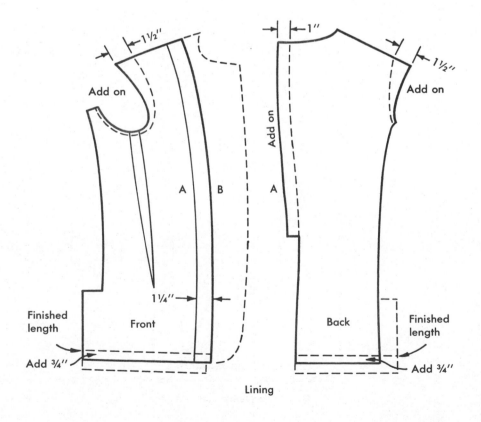

Lining

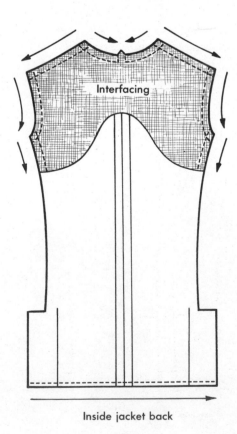

Inside jacket back

Jacket back unit

For pleat extension, mark pleat line on one piece only. Pin line with two pins through both layers of fabric. Turn back and press top piece of fabric on pin line. Remove pins. Turn over to second piece of fabric. Fold pleat line and key to first piece. Press on second side. See photographs on pages 164–165.

Stitch center back seam on seamline in direction shown, and press open in same direction. Beginning 1¼ inches from neck edge, trim seam away diagonally to ⅛ inch to eliminate bulk at seamline.

Place interfacing next to underside of jacket. Staystitch interfacing to jacket, in direction shown, just outside seamline.

Staystitch lower edge ¼ inch from edge.

Interfacing

Inside jacket back

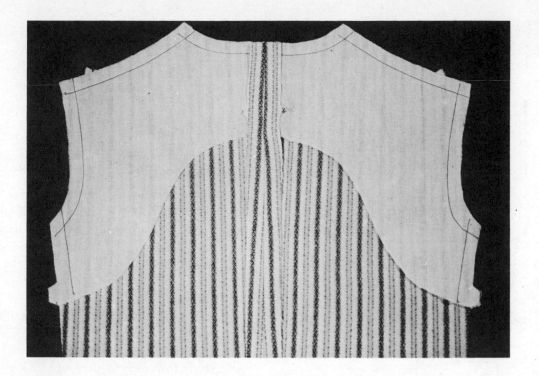

It is advisable to staystitch interfacing in center back seam for support in lightweight fabrics.

Open out pleat extensions. Using automatic hem gauge, turn up hem 1½ inches and press. Turn back pleat extensions again and press to restore pleat line.

Trim center back seam in hem to ¼ inch, going ⅛ inch beyond fold of hem.

Place bias muslin (cut 2½ inches wide) in hem 1 inch from lower edge of jacket. Extend it ½ inch beyond crease line of pleats. At points A, cut away muslin diagonally to remove it ¼ inch from each side of corner of jacket. Stitch muslin in place ¼ inch from edge at sides and lower edge of muslin. Stitching will be ¼ inch from fold lines on jacket.

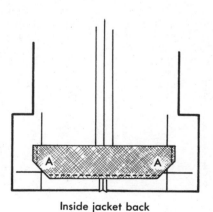

Inside jacket back

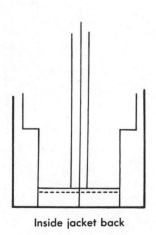

Inside jacket back

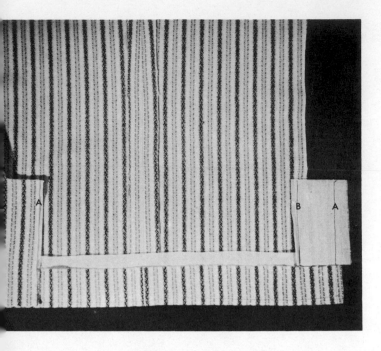

In this lightweight jacket, a straight strip of muslin was also stitched in place at A and B to support open pleat.

Finally, press hem and pleat extensions in place again to restore them.

Jacket front unit

Stitch dart in jacket front and press toward side. At A, slash dart on fold line for a depth of 1½ inches. Trim under edge away diagonally to ⅛ inch at armhole and upper edge to ⅜ inch. Staystitch lower edge of jacket ¼ inch from edge.

On upper left front, make regulation welt pocket, following directions on pages 174–176, *BMCC*. It is important to remember that the grain of the under pocket piece must match the grain of the jacket front. See jackets in photographs on page 184.

On both jacket fronts, sew on patch pockets, following directions on pages 176–177, *BMCC*. The top pocket is made through the interfacing, but the bottom ones never are.

Following are two additional techniques you may choose to use in making the patch pockets.

In some fabrics, the pocket will be stronger with interfacing across the top.

Press hem allowance in place at top of pocket. Cut a lengthwise strip of hair canvas ¾ inch deeper (A) than width of hem and ¾ inch narrower (B and B) than width of pocket. Place inside hem up to fold line. Open out hem, and stitch interfacing to hem ¼ inch below fold (C). When lining is stitched on pocket edge (D), stitching will catch interfacing here, also. Follow the directions at the top of page 177, *BMCC*, for sewing on lining.

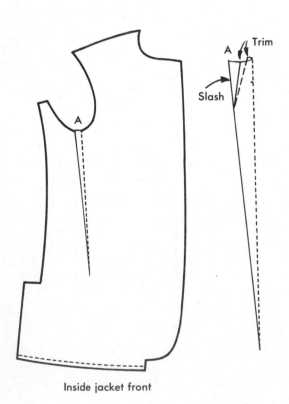

Inside jacket front

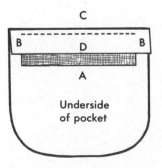

Underside of pocket

Inside the right front patch pocket, a little inside pocket may be made for keys or change!

Cut a piece of lining as indicated.

With right sides together, fold in half on lengthwise grain, and stitch ¼-inch side seams. Turn and press.

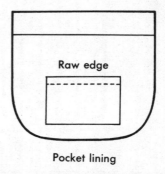

Pocket lining

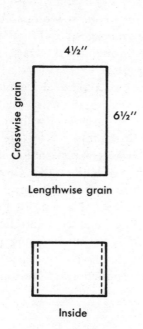

4½″

Crosswise grain

6½″

Lengthwise grain

Inside

Outside of inner pocket

Stitch together raw edges ¼ inch from edge.

Stitch raw edge to patch pocket lining.

Turn up pocket and press. Stitch sides to patch pocket lining close to edge and diagonally at top for reinforcement.

Stitch together pocket piece and lining *precisely* ¼ inch from raw edges. Sew on pockets as directed in photographs on page 177, *BMCC*.

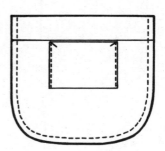

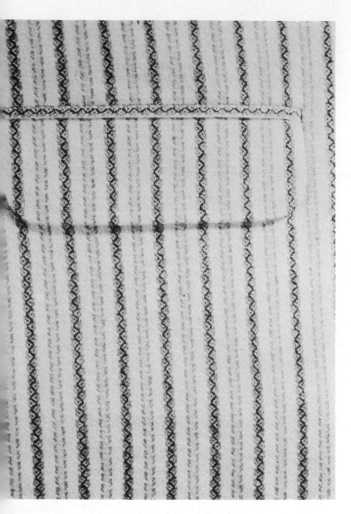

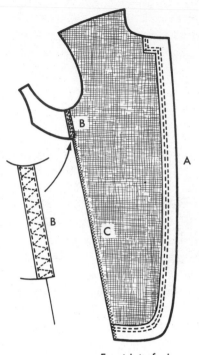

Front interfacing

The second jacket photographed on page 184 has flap pockets instead of patch pockets at the bottom of the jacket.

They were made exactly like the directions on pages 172–174, *BMCC*, except that the lengthwise folded strip across the top was sewn to the flap before the flap was applied to the pocket piece.

Using automatic hem gauge, turn up hem of jacket 1½ inches and press.

Jacket front interfacing unit

Match outer edge of the muslin strip with the outer edge of hair canvas (A). Place muslin strip on opposite side of hair canvas from center front and buttonhole markings.

Stitch muslin to hair canvas ⅞ inch from edge; stitch again near inside edge of muslin strip for reinforcement; press if necessary.

Butt edges of muslin and hair canvas interfacing at B and apply a piece of seam tape. Stitch at both edges of tape and zigzag back and forth to keep flat the butt edges of interfacing.

At C, use zigzag stitch on edge of hair canvas to control the off grain edge, or a staystitch ¼ inch from edge may be used.

193

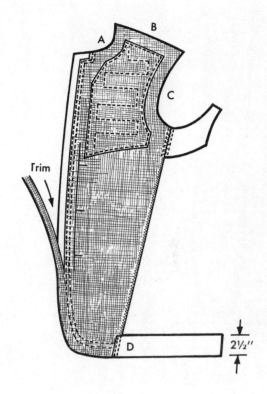

Trim away hair canvas to first row of stitching ⅞ inch from edge.

Place reinforcement piece on opposite side of interfacing from muslin strip. It will be on the same side as the center front and buttonhole markings. Reinforcement piece will be ⅞ inch from A, B, and C edges. Stitch around this section ¼ inch from all edges. Quilt a design at machine on true grain of section for added strength. If Pelomite is used, just press in place with hot iron.

Overlap bias muslin (cut 2½ inches wide) on hair canvas interfacing for ¼ inch (D). Stitch at edge of muslin and hair canvas to hold securely.

Place interfacing to underside of jacket front, muslin strip facing jacket. At A, cut away interfacing diagonally at corner ¼ inch beyond seamline to eliminate bulk.

Staystitch interfacing to jacket in direction shown just outside seamline. On front edge B, staystitch exactly on seamline up to center front line.

Stitch bias muslin in place at C about ¼ inch from lower edge of muslin. Stitching will be about ¼ inch from hem line. Muslin extends to raw edge of jacket side front (D).

In contrasting thread, baste-stitch center front and crosswise lines for buttonholes from interfacing through to outside of garment. Baste-stitch the crosswise lines before doing the lengthwise ones.

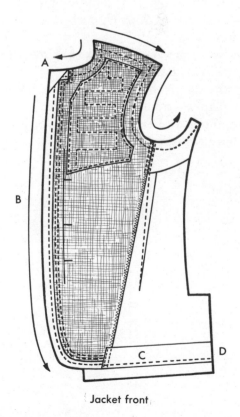

Jacket front

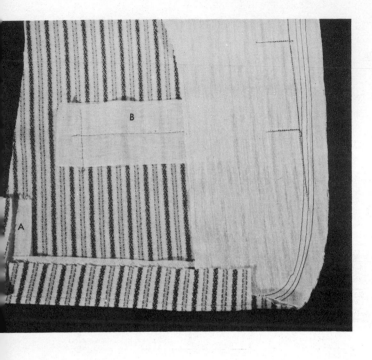

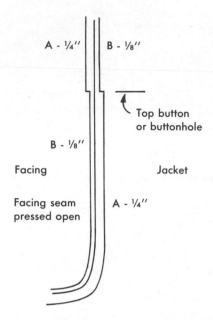

A - ¼"   B - ⅛"

← Top button
or buttonhole

B - ⅛"

Facing   Jacket

Facing seam   A - ¼"
pressed open

In this lightweight jacket, just as you learned to do with the back unit, a straight strip of muslin is stitched in place at A to support the side pleat.

Flap pockets are always reinforced with a lengthwise muslin strip (B).

Front facing unit

Staystitch neck (A) and inside (B) edges of facing on seamline. Staystitch lower edge (C) ¼ inch from edge for a distance of 2½ inches from bottom up.

Right sides together, place facing on jacket front, keying all edges with perfection.

With interfacing side up, stitch facings to jacket from lower edge of facing to center front line. Sew through staystitching on seamline and *favor* facings above markings for top buttonhole and button. Use short stitches for ½ inch on either side of corner.

At neckline, snip facing and jacket to seamline at center front.

Trim away muslin strip to seamline and press open facing seam on edge presser.

Stagger facing seam in trimming. The upper seam (A) is trimmed down to ¼ inch and the under seam (B) to ⅛ inch. At the top buttonhole and button, the ¼-inch seam will change to ⅛ inch, and the ⅛-inch seam to ¼ inch, because the facing reverses and becomes the topside. Trim away seam allowance in corner to within a few threads of seamline.

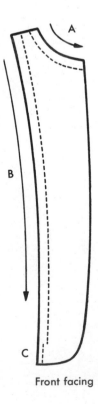

Front facing

Turn facing to inside and set the edge, pressing with underside *on top*. From the top buttonhole and button coming down, the underside will be the facing. From the top buttonhole and button, going up to the center front, the underside will be the garment.

After the edge has been set in pressing, turn and have jacket right side up. Press again and flatten facings, pockets, and hem with pounding block. Flatten lower edge of jacket back, also.

This photograph shows the finished front facing unit at the neckline.

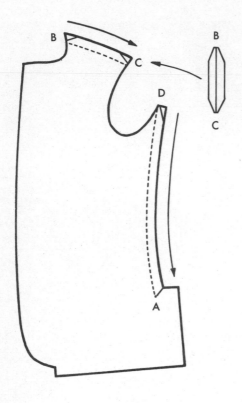

Jacket unit

Stitch shoulder and side seams in direction shown. Clip diagonally ⅝ inch at A for pleat extension. Press open seams in the same direction they were stitched.

Leave interfacing in shoulder seam, but trim out of side seam near seamline. At B, C, and D, beginning 1¼ inches from edge, trim seam away diagonally to ⅛ inch to eliminate bulk at seamline.

Collar unit

Staystitch with a longer stitch the top collar ⅝ inch from the edge in contrasting thread. At A and B edges, turn back seam allowance on staystitching, and press with perfection. At C corners, cut away bulk of seam, and whip raw edges together on a diagonal line.

The under collar is cut of lightweight felt, hard flannel, or other similar fabric; the interfacing has been cut of hair canvas and on the bias, as was the under collar. Staystitch ¾ inch from A, B, and D edges. Join center seam on seamline, trim interfacing away to seamline, press open seam, and topstitch both sides of seamline. Trim away under collar seam to topstitching.

Mark a quilting pattern on true grain of hair canvas for under collar.

Stitch quilting pattern at machine through felt and hair canvas. Trim hair canvas up to staystitching ¾ inch from edge at all outside edges, and trim felt ⅝ inch from edge.

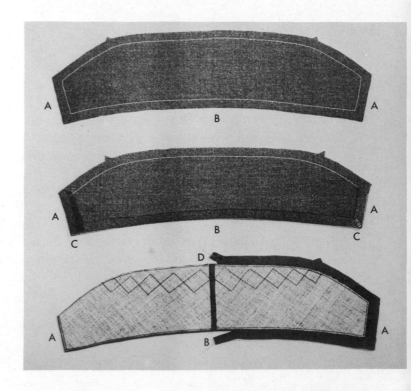

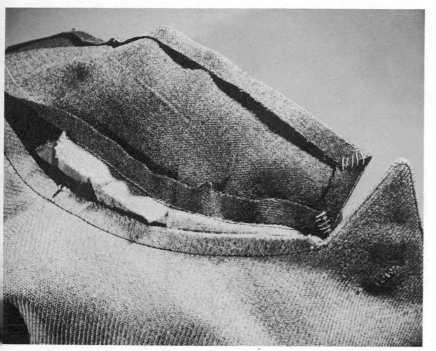

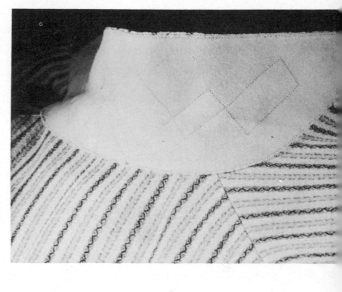

Sew top collar to front facings from center front to ⅝ inch from edge of facings at shoulder seam. Keep facing side up and sew through staystitching on seamline. Press open seam. Trim facing seam to ¼ inch and top collar seam to ⅜ inch as far up as shoulder seam of facings. By hand, loosely stitch front neckline edges of jacket to top collar seam.

Whipstitch the under collar to the top collar. Remove staystitching in contrasting thread from the top collar. The top collar is still free across the back from one edge of the front facing to the other.

Press the collar into a sharp crease at the fold line. It should fit the shirt collar with perfection. Use a cushion to shape the collar, and shrink, mold, and press the fold line. The crease extends ½ inch below the seamline of the collar into the lapel.

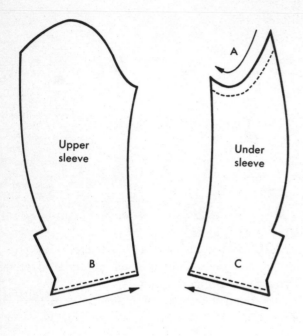

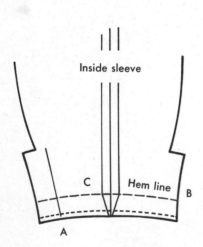

Sleeve unit

Staystitch A edge on seamline and B and C ¼ inch from edges in direction shown.

Stitch sleeve seam (D) in direction shown. The grain on this sleeve is such that the seam should be stitched from the bottom up. Press open seam, and at E, beginning 1¼ inches from edge, trim seam away diagonally to ⅛ inch to eliminate bulk at seamline. Trim seam at F to ¼ inch for depth of hem (1½ inches).

Cut a piece of hair canvas on the bias 3½ inches wide and long enough to extend ¼ inch beyond the fold line at A and all the way out to the raw edge at B. It will extend ½ inch beyond hem line (1½ inches from lower edge) at C.

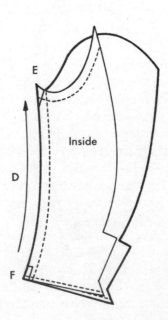

Stitch hair canvas in place ⅛ inch from edge at D, E, and F. Turn up and press hem of sleeve 1½ inches from lower edge, and turn back and press pleat extension on fold line.

Stitch second sleeve seam from armhole down, including pleat and hem, and using small stitches at corner G.

At G, clip diagonally to stitching in corner on under sleeve section.

At H, stitch all layers together ⅛ inch from edge, and bevel away with scissors all the under layers of fabric up to ⅛-inch stitching.

Press open sleeve seam and final-press pleat with perfection. At I, trim seam away diagonally.

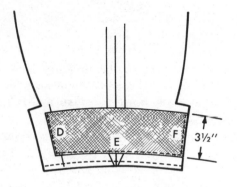

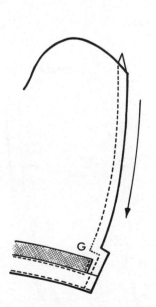

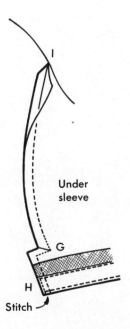

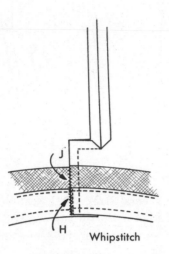

Whipstitch

At H, whipstitch pleat edge to hem, only catching hem in stitching. Stitch down to bottom of hem and back up to top of it to hold edge securely. Whipstitch will have the appearance of a cross-stitch. At J, whipstitch pleat extension to hair canvas only. Catch-stitch hair canvas to second sleeve seam. See sketch with the lining on page 202.

Sew on 2, 3, or 4 buttons. They should start 1¼ inches from lower edge of sleeve and ½ inch from pleat line. They should never extend above the pleat line. It is not necessary to make a shank as you learned on pages 84–85, *BMCC*.

Draw buttonhole twist through the beeswax shown in the holder in the left of the photograph. This waxing prevents the thread from knotting and twisting. Used thread double in the needle, but separate it to make the chain stitch, as shown in the photograph. It is a false buttonhole and gives a nice, custom-looking touch.

Stitch sleeve seams of lining in same direction as jacket sleeve seams were stitched. Press open.

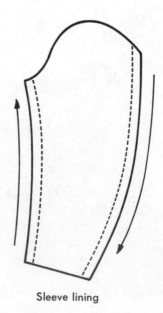

Sleeve lining

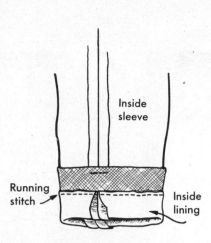

Inside
sleeve

Running
stitch

Inside
lining

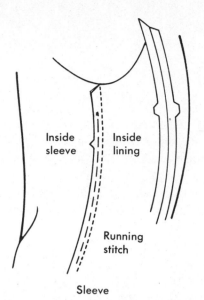

Inside
sleeve

Inside
lining

Running
stitch

Sleeve

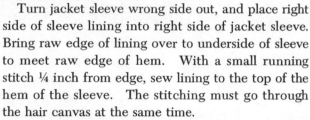

Turn jacket sleeve wrong side out, and place right side of sleeve lining into right side of jacket sleeve. Bring raw edge of lining over to underside of sleeve to meet raw edge of hem. With a small running stitch ¼ inch from edge, sew lining to the top of the hem of the sleeve. The stitching must go through the hair canvas at the same time.

Bring the lining out of the jacket sleeve. With wrong sides together, catch seams of lining and jacket. Beginning about 2 inches below the armhole, attach one side of lining seam to corresponding side of sleeve seam with a single thread and a long, easy running stitch through the middle of the seam allowance. Fasten the thread about 3 inches above bias hair can-

vas. Make several back stitches at the beginning (even with a knot in the thread) and at the end of the stitching to secure the thread in fabric. Always favor the lining. Repeat for second seam.

Place hand in lining sleeve and turn on top of the jacket sleeve, turning the lining right side out. Press down excess lining at lower edge into what is known as a take-up tuck; this is necessary to prevent elbows from wearing through. See photograph on page 165, *BMCC*.

From one notch to the other, do off-grain stitching (page 157, *BMCC*) to prepare the sleeve cap to fit into the armhole.

Sew sleeve into armhole (page 157, *BMCC*). Do not trim any of the seam allowance or any interfacing from armhole until stitching in lining. For added strength underarm, you may stitch the seam a second time below the notches on the seamline.

To maintain a firm line on the cap of the sleeve, and to round out the ease, cut a piece of jacket fabric on true bias 3½ inches wide and 10 inches long. This length will vary with the size of the jacket. Fold and press lengthwise, bringing one raw edge within ½ inch of the other. With shorter raw edge (A) turned up, place folded edge of bias (B) ¼ inch from raw edge of armhole seam. Sew by hand as close as possible to the seamline of sleeve.

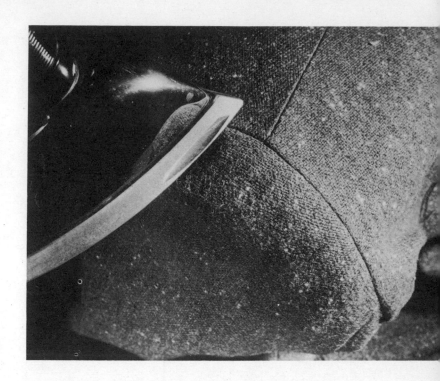

The armhole seam turns into the sleeve. From one notch to the other over the top of the sleeve, press seam flat on a cushion. Roll the iron out on top of the sleeve in pressing. Do not press seam underarm.

Men's garments have a smooth, round line over the cap of the sleeve, but women's garments have a soft, rolled one.

Match raw edges and key armhole of lining to armhole of sleeve at notches, underarm seam, and shoulder seam. With a loose running stitch, tack sleeve cap of lining to that of jacket, easing in lining fullness over cap. Below notches, take small running stitch close to seam. Trim seam (four thicknesses) to ¼ inch from notch to notch under the arm (see top photo, page 166, *BMCC*).

If shoulder shapes are needed to obtain a normal shoulder line, shoulder shapes are used in men's garments the same as women's. (Follow directions on pages 158–59, *BMCC*, for making them.) They should be sewn into the jacket at this time, before the body of the lining is attached. Key the shoulder seam of the shape to the shoulder seam of the jacket; also, key raw edge of shoulder shape at armhole to raw edge of armhole seam of jacket. With a running stitch, attach pad to jacket at shoulder and armhole seams. Catch edges of pad to muslin interfacing in back and hair canvas in front.

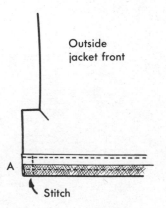

Side pleats unit

At jacket side front, turn hem to outside on fold line, right sides together. Stitch ⅝-inch seam to top of hem (A). Press open seam, trim to ¼ inch, round off fabric in corner, turn right side out, and press.

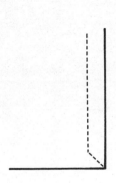

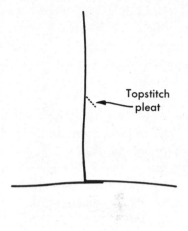

Topstitch
pleat

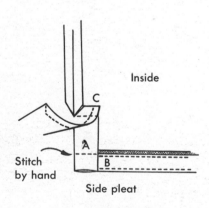

Inside

C

Stitch
by hand

A

B

Side pleat

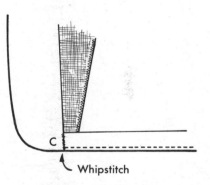

C

Whipstitch

Topstitch ¼ inch from edge, going diagonally at lower corner.

At jacket side back, stitch facing by hand at A to hem only. Whipstitch B the same as pleat extension on sleeve. Lap raw edges of front and back pleat extensions to meet at C, and stitch together on machine.

Topstitch pleat at slight angle for ¾ inch from pleat edge. The two pleats at the sides of a jacket are preferred over one at the center back, and most jackets have two instead of one.

Finish lower edge of front facing (C) the same as pleat extension on the sleeve.

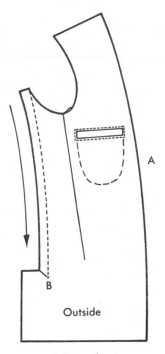

A

B

Outside

Left lining front

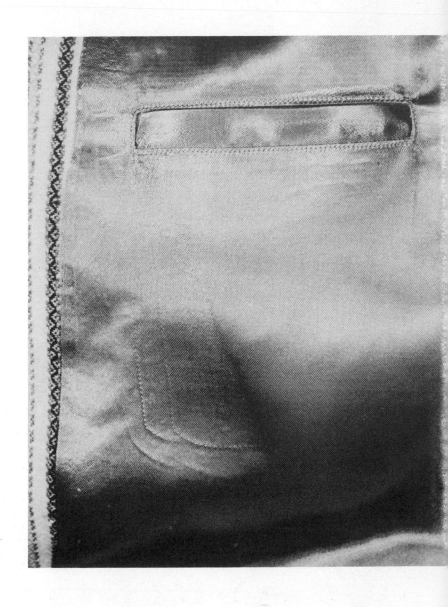

Lining unit

On left side of lining front, and beginning 1¼ inches from front raw edge (A), make modified welt pocket (pages 114–115, *BMCC*). Sew side seams of lining fronts and back in direction shown to marking for pleat extension. On lining fronts, clip diagonally to stitching at pleat opening (B).

This photograph shows the modified welt pocket in the completed front lining. Press open seams. Sew underarm darts and press toward front; jacket darts have been pressed toward back. Staystitch back neckline on seamline.

The center back pleat is stitched the same as the woman's jacket on page 165, *BMCC*. Catch-stitch in position at the waistline.

Place right side of lining to right side of front facings, and with facing side up, sew lining to both facing edges with a ⅝-inch seam to 2 inches above top of jacket hem. Stitch one thread inside staystitching on facings on seamline. Use the longest stitch at the machine unless it puckers. If so, shorten stitch slightly.

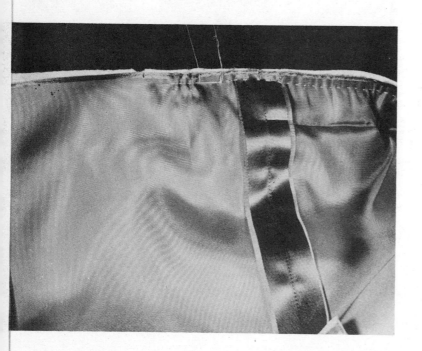

Bring the raw edge of the lining over to the underside of jacket hem to meet raw edge of hem (same as sleeves were done). With a small running stitch ¼ inch from raw edge, sew lining to hem, catching front interfacing and center back seam.

Turn lining to inside of jacket and with wrong sides together, beginning 2 inches below armhole, catch underarm seams of lining and jacket, same as sleeve lining seams (page 202).

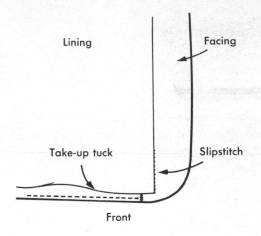

Press down excess lining at lower edge of front facings. This is known as a take-up tuck, and is necessary for longer wear on the lining. Complete lower edge of front facings with slipstitching and include take-up tuck to hold in place. Press lining flat at front facings.

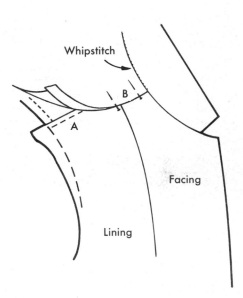

Key raw shoulder edge of lining front to raw edge of back half of jacket shoulder seam (A). Attach with an easy running stitch.

Turn under seam allowance on the shoulders of lining back and lap ⅝ inch over lining front; pin in place (B).

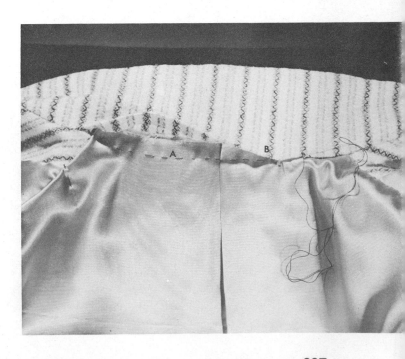

Match neck edge of lining to seamline of jacket, and attach with an easy running stitch (A). Turn under top collar ⅝ inch and bring ⅝ inch over lining back (B); pin in place. Whipstitch top collar and back shoulders in place. Make small stitches with heavy-duty thread or buttonhole twist.

Whipstitch lining in place around the armholes the same as a woman's garment (bottom of page 167, *BMCC*).

This technique for finishing the body of the lining over the sleeve lining is not the usual method found in ready-to-wear. However, it gives longer wear; there is no pull on the stitches; it is easier to do, and is a repeat performance of the method in women's tailoring.

To complete side pleats, turn in lining front to ¼ inch from finished edge of pleat (A); slipstitch. Finish take-up tuck same as front edge.

Slash into back lining at an angle until back lining can be turned under to meet edge of front pleat with perfection (B). Cut away excess lining. Turn under and slipstitch top and side to complete pleat extension. Finish take-up tuck same as front edge. Press take-up tuck in place at entire lower edge of jacket.

Make keyhole buttonholes in left front and regular buttonhole on left lapel, if desired. They may be done with a buttonhole attachment, and then embroidered over that by hand with a buttonhole stitch, using buttonhole twist. Sew on buttons with a good shank on right front (pages 84–85, *BMCC*).

Remove baste-stitching at pleat in center back of lining. Press lining at edge of top collar, but not across shoulders or around armholes. If the outer fabric has become wrinkled from the handling as the lining was being attached, some light final-pressing may again be necessary on the jacket.

Chapter 14

Trimming details for the quality look

The chapter in our first book on trimming details was so well received that we are adding one to this book, also.

The principle in the Bishop method has been that you work toward learning to express yourself by adding a trimming detail when it will help to obtain a quality look. Do not be a slave to a pattern, but add the right personal touch to your garment, as a designer will do.

In seeking the right trimming detail, you will often get a hint from either the fabric or the style of the garment. Many times you will get an idea from a fabric, especially a design idea, before ever cutting it out. Other times an idea will come to you as you work on the garment; or playing with scraps may give you an idea in seeking the right detail for the quality look.

However, as much as we have stressed trimming details in these two chapters, we would also like to emphasize that not every garment needs trimming. The lines of the garment may be impressive, or the fabric may be individual. The garment may then stand completely on its own.

1. The rose was made following the directions with the next photograph. The leaves were made on the bias, as shown on page 223. Instead of a buttonhole, the jacket is fastened with a silk-covered snap, sewn on the facing under the rose.

The rose would be attractive many other places—at the waistline, near the hemline in clusters, grouped at a neckline, etc.

Embroider French knots in a square of fabric (A). Cut a circle of cardboard, place a mound of cotton on top, and cut square (A) to fit around circle; stitch by hand. The cotton gives a raised effect in the center. Fold true bias (B) right sides out, and stitch near raw edges. Lap small pieces of bias and tack around circle shown at C. The last piece of bias (D) is sewn from the underside; then, it rolls up to cover raw edges of circle and bias.

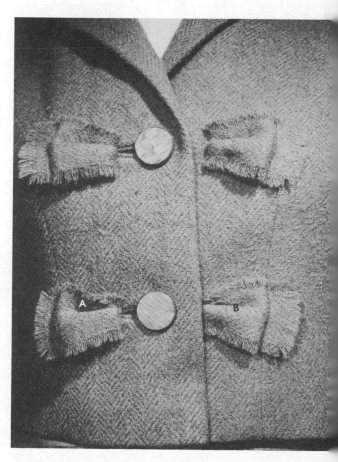

2. When a suit or dress is completed, these bows may be added. The half of the bow at A is placed at the edge of the buttonhole. The other half of the bow at B is placed on the other side just at the edge of the right front of the jacket. See directions with the sketch that follows.

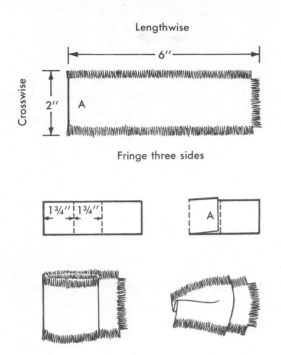

Lengthwise

6"

Crosswise

2"

A

Fringe three sides

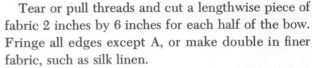

1¾" 1¾"

A

Tear or pull threads and cut a lengthwise piece of fabric 2 inches by 6 inches for each half of the bow. Fringe all edges except A, or make double in finer fabric, such as silk linen.

Make a double turn at A edge for a depth of 1¾ inches each time. Fold a pleat at inside edge of bow and invisibly stitch to hold in place. Press this edge only. Hand sew bow in place on garment.

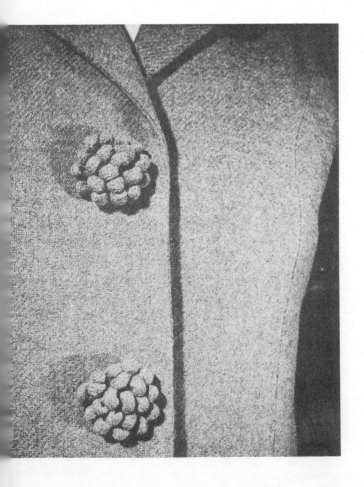

3. These flower buttons may be the real touch of drama on a jacket or dress. Made of self-fabric, they are sewn on the jacket, and silk-covered snaps (page 214) are used to fasten the garment on the underside. However, the buttons could be used equally well with buttonholes. To make these buttons, see directions with the next photograph.

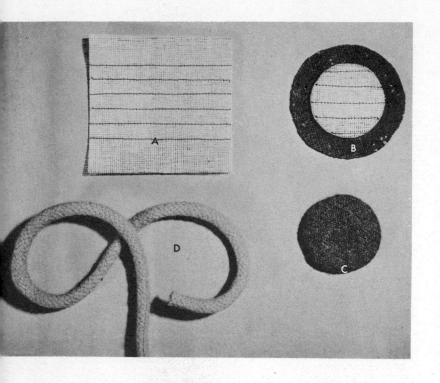

Sew together several layers of crinoline or buckram (A). Then, cut circle desired size, and place in the middle of a self-fabric circle (B). Run a baste-stitch around fabric circle ¼ inch from edge, and draw up basting; fasten circle neatly with hand stitches (C). Make bias strip (D) as you learned on page 193, *The Bishop Method of Clothing Construction.** Press open on wooden dowel, shown at A in next photograph. Beginning at center, on the smooth side of the circle, fasten an end of the bias strip. Make a loop and tack it to the circle; continue until the circle is covered with the loops.

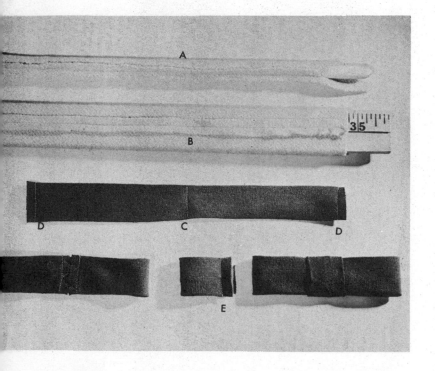

4. Many times, there is nothing that will add so much to a dress as a pretty bow placed with an artist's touch here or there.

The A was just added to the photograph to show how narrow bias can be pressed open on a dowel. The wider bias used for the bow is shown at B, being pressed open on a yardstick. The seam will be placed to the center of the underside.

The row of stitching at C identifies the center for the bow to which the raw edges at D are placed and stitched flat.

The center of the bow is also seamed (E), pressed open, turned right side out, and slid in place at the center of the bow.

These basic techniques will help you in making any bow.

*Referred to hereafter as *BMCC*.

5. The pocket flaps are the important detail on this jacket; so the buttons and buttonholes are few in number, and plain, covered buttons were used.

The fringes on A flaps are a different color from the fringes on B flaps, because the A ones were made from a lengthwise strip, while the B ones were made from a crosswise strip. At the lower edges (C), the fabric was folded, and the threads were pulled from the top layer only to form the fringed look. That way, the fringe will not fray at the lower edge.

6. This detail would be smart-looking on the front of dresses and blouses as well as suits. Silk snaps were used to fasten this jacket on the underside, but buttonholes could easily be made under the flaps.

After these square flaps were fringed, two were sewn together by hand near the inside edge of the fringe to give each one more character than a single layer of wool. Then, the flaps were tacked to the jacket with invisible hand stitches at the upper corners.

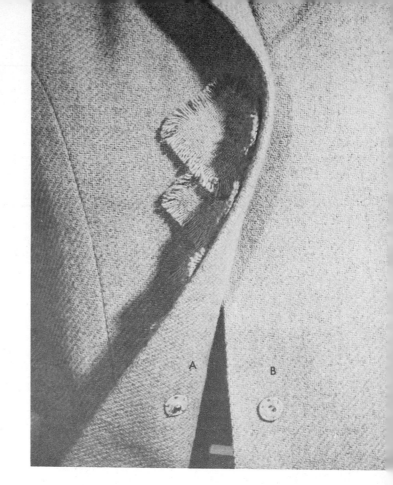

7. Covering snaps (A and B) with matching lining fabric is an important step toward having quality looking clothes. Use snaps approximately ½ inch in diameter. Follow directions below.

The circle (A) is usually the size of a small spool of thread. Punch a hole with a stiletto and place lining over snap (B). Sew around circle of lining; draw up thread and fasten neatly on underside of snap.

Place second circle over second half of snap. Close the snap, and while the snap is together, sew around second circle of lining; draw up thread and fasten neatly on underside of snap.

C and D show the completed snap before it is sewn on the garment. Feel with the needle for holes in snaps, and be certain to sew snap to garment through holes of snap.

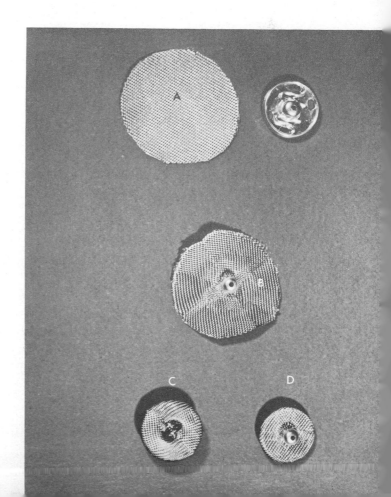

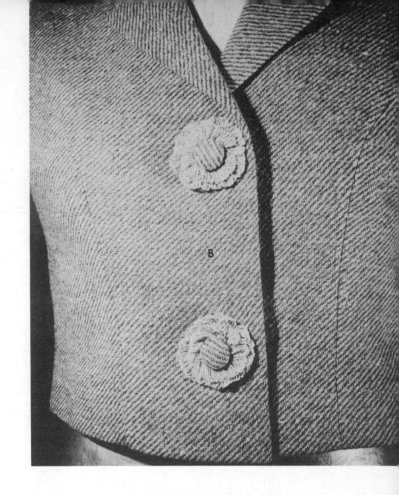

8. Buttons with fringe of matching fabric will be a conversation piece on a suit. The ones on A jacket are used with buttonholes, but the ones on B jacket have the silk covered snaps on the underside. It would be overpowering to have three of this detail on a short jacket—two are better in design. One should be placed at the bustline.

Two circles of folded fringe are sewn under the covered button. You can see that the fringe in A, which is crosswise, would be a different color from B, which is lengthwise of the fabric. Use the color that is most flattering to your fabric.

Fold and press fringe and sew through seam allowance with a running stitch; draw fringe to form a circle; press flat. The second circle is made smaller than the first.

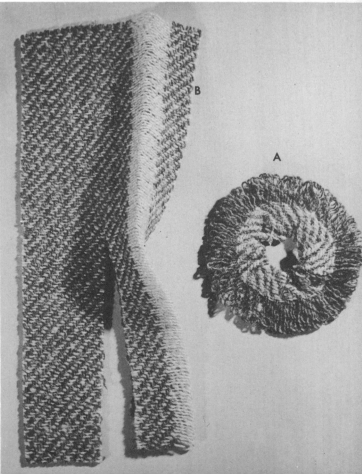

9. When a dress, blouse, jacket, or coat is completed, these interesting tabs can be slipped through the buttonhole, and hand sewn in place. They are made double, and slashed on the underside (page 121, *BMCC*) to turn and press.

We are showing the pattern for the tabs without any seam allowance. However, they may be made smaller, depending upon the type of garment and the number used.

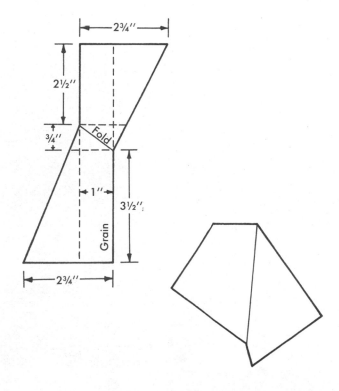

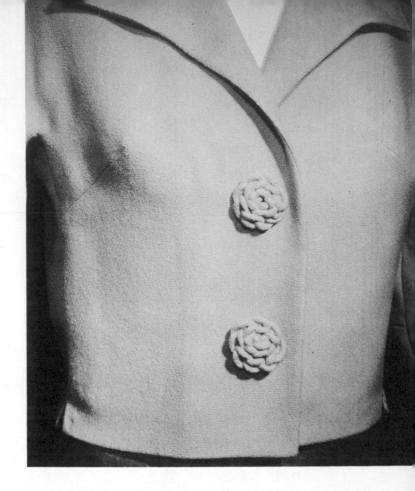

10. These crocheted buttons will add charm to any tailored garment. They are sewn on this jacket, and silk snaps were used on the underside for fastening. The pattern is the same as the suit on page 213.

To make the crocheted buttons, take the end of the bias and form a loop; stitch by hand. Then, make six single crochet stitches; join to first permanent loop. Continue with double crochet stitches around the first six single ones. Pull bias through last stitch and stitch by hand on underside. Pull starting loop up to make center of button (A).

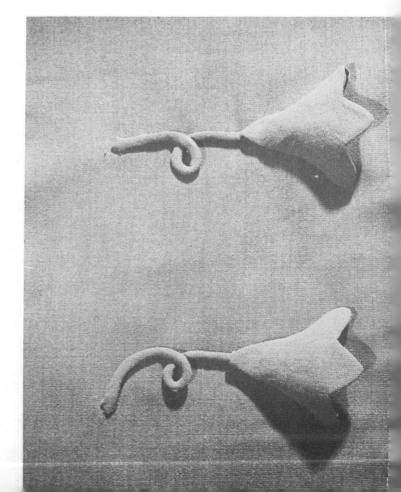

11. These tabs will be unusual used on the lower edge of a jacket, bolero, toppette, or skirt. They can also be sewn into the seam of a collar. Do not make them in heavy fabric.

Fold raw edges to meet in center of tab (A). Then, fold tab in half and stitch ends together (B). The tabs are ready to insert into a seam or on an edge with a facing.

12. These tulips give a designer's touch and can be used over the buttonhole on a suit of linen, silk linen, lightweight woolen, etc. A machine buttonhole is all that is necessary, because it will not show. The directions are given with the next photograph.

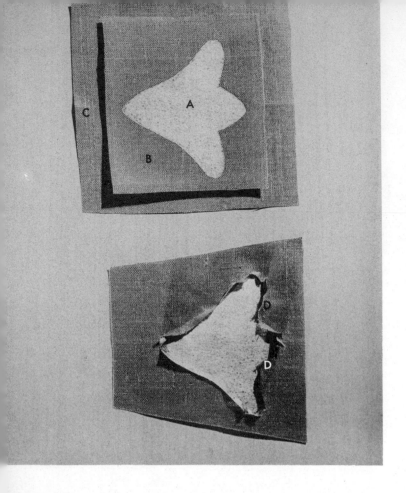

Cut a piece of Pelomite the desired size of finished tulip (A). Press on a square of fabric so that the lengthwise grain of fabric goes from top to bottom of tulip (B). Place a second square below the first one (C) on identical lengthwise grain.

Stitch around edge of Pelomite with smallest stitch machine will make. Trim square B to ⅛-inch seam allowance. Clip to stitching at points of tulip (D and D). Press seam up. Then, trim under-square C to ⅛-inch seam allowance and clip to stitching at points of tulip (D and D). In center of tulip, make a slash on underside to turn right side out and press.

When tacking tulip to garment, allow for sufficient space to cover button.

Bias (page 193, *BMCC*) is slipped under lower edge of tulip when being invisibly sewn to garment.

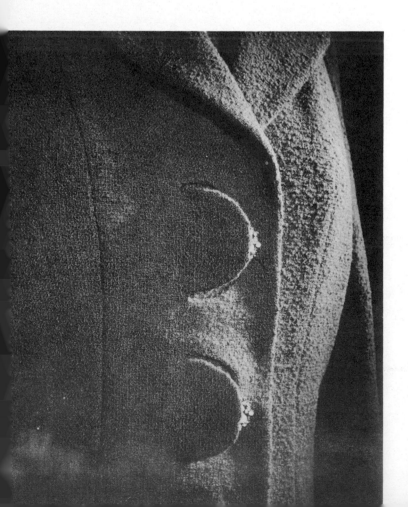

13. The half-circle button closure is extremely smart on a jacket with jeweled or special-interest buttons.

Trimming details for the quality look

Mark position of half-circle opening on underside of garment.   Cut a facing patch on identical grain of jacket, and place on jacket, right sides together.   Using a small stitch, sew a few threads either side of the half-circle marking.

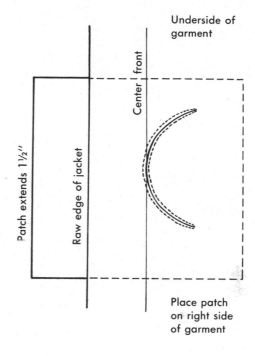

Cut through center of stitching.   Pull facing patch through slit to underside of garment.   Press entire area.   Next, fold facing patch back on half circle at fold line marked on the sketch.   Press carefully on grain.   Stitch suitable interfacing or underlining for garment on underside.

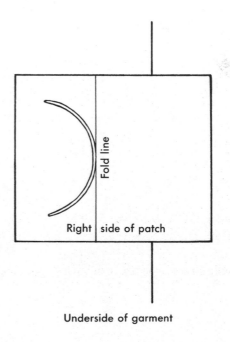

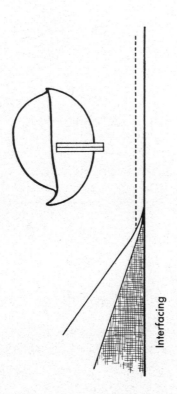

Make a tailored buttonhole through facing patch and interfacing.   Proceed with completing jacket by putting on facings.   Use a button with special interest.

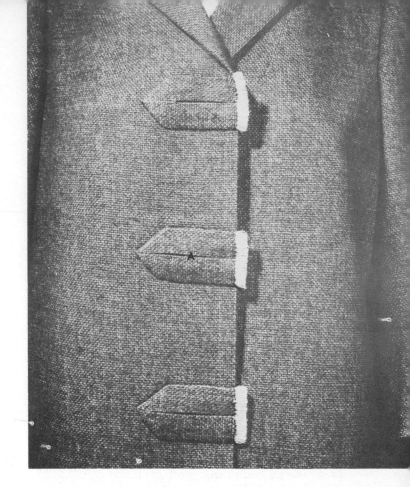

14. This intriguing detail could be used on any garment. The straps may cover the buttons and buttonholes, and you would reach inside at the opening (A) to open and close the garment.

However, if it is desired to display the buttons, they may be brought on top of the strap, as in this photograph.

To make straps, see directions with sketch on next page.

Fold

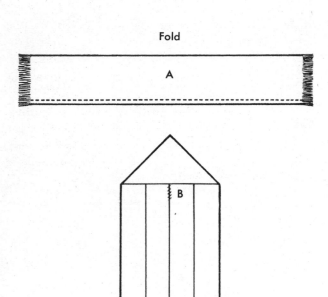

To make straps, fold a lengthwise strip of fabric in half (A), right sides together, and make a ¼-inch seam. Fringe both ends. Press seam open on dowel (page 212), turn strap right side out, and press flat with seam in the center on the underside.

Press to form a perfect mitered corner. Invisibly stitch strap together at B and B. Slipstitch in place on garment.

15. Starting at both shoulders, this versatile detail makes an attractive, self-trim on this peau de soie dinner dress. Note how it was used to give a finish at the waistline and on the lower edge of the sleeves. The next photograph describes the steps in making it.

To have a rounded and finished end, as shown at A in the photograph, fold true bias (B), right sides together. Then, fold lengthwise line in half, and stitch a mitered line, ending ¼ inch from lower fold (C).

Trim away excess fabric at miter, and pull two bias strips right side out.

Stitch the four raw edges together (D) in a narrow seam, holding true bias with perfection at fold line. In some fabrics, you may pull slightly as you stitch.

16. Two pockets were used for the trim on this dress. The leaves were made lengthwise following the directions on the bottom of page 121, *BMCC*. They were folded into a small tuck at the top to give a full look and were sewn by hand to the pockets.

The two pieces of bias that are tied at the waistline are just long enough to extend into the upper pocket.

The four pieces of bias below the pocket are sewn in place under the lower pocket and lower edge of skirt.

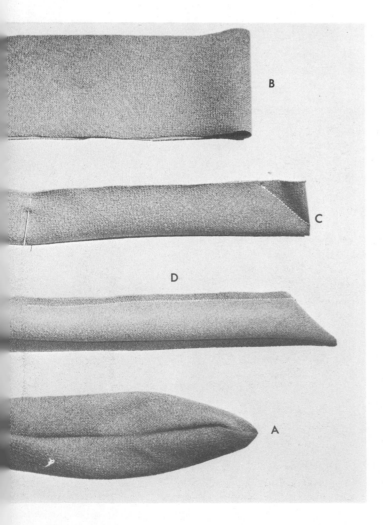

223

17. A basic dress gives you a very wide choice of detail. Large pockets continue in the fashion picture; these double ones are made on lengthwise grain. The fringe is double, and was made in the body of the fabric before the hem was turned. The pockets are placed and tacked over the belt.

18. Sequins with small crystal beads sewn in the center of them make a timeless detail and attractive outline for a print. They were used on the collar and bow of this dinner dress. The color of the sequins should accent the most becoming color in the print.

The finish on the bottom of the sleeve is the simulated cuff shown on page 76, *BMCC*.

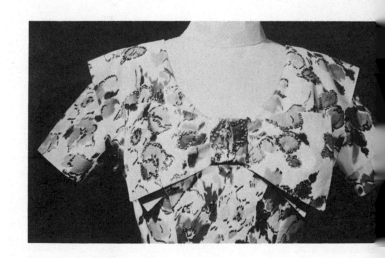

19. A folded piece of bias is stitched on the waist-line seam of this dress and then turned down. It not only makes an attractive finish, but is more slenderizing and more flattering on many figures than a belt.

20. (*bottom left*) A bias bow can be placed in many attractive ways on a garment for a quality looking detail.

The bias finish at the waistline is the same as that on the previous dress.

21. (*bottom right*) Appliqués in an unusual design can add much quality to a basic dress.

Cut design from Pelomite and press on lengthwise piece of fabric. Trim fabric to within ¼ inch of the edge of the Pelomite, clip into corners, and press seam to underside.

Place appliqué on center dress front a little higher above than below the belt-line; pin in place. Invisibly slip-stitch appliqué to dress. For a wool or dull crepe dress, make design of matching satin or velvet. A shade darker or lighter than the dress color is also attractive. Self fabric is always a smart choice, too.

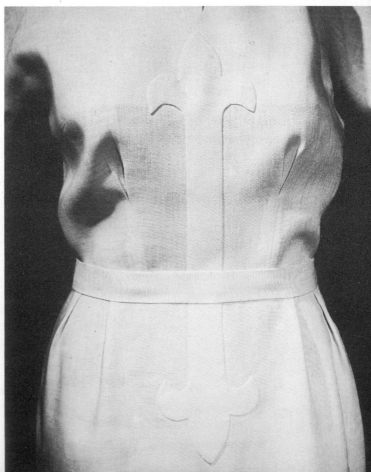

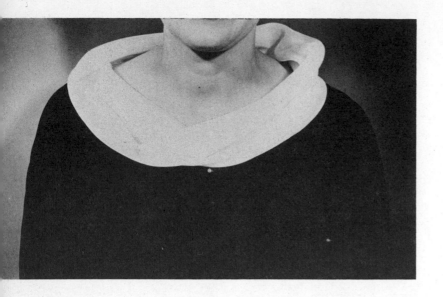

22. The "smoke ring" gives an added soft touch to a plain bateau or scoop neckline. This one was made of silk chiffon, but other fabric that has a draping quality (such as Paisley silk print) is desirable, too. The directions are given with the following sketches.

Cut a piece of fabric on true bias, 31 inches long and 14½ inches deep. Ends A are straight of grain.

Fold fabric lengthwise, right sides together, and sew a ¼-inch seam, beginning and ending ¼ inch from each end (B). Press open seam. Turn right side out. Place together right sides of straight-of-grain edges (A in first sketch) and sew a ¼-inch seam. Press open seam, and pull to inside.

Right sides together, sew remaining bias edges together, leaving a small opening to turn right side out.

Turn in raw edges and slipstitch opening by hand. Press smoke ring. If ring is too large for neckline, shorten length several inches.

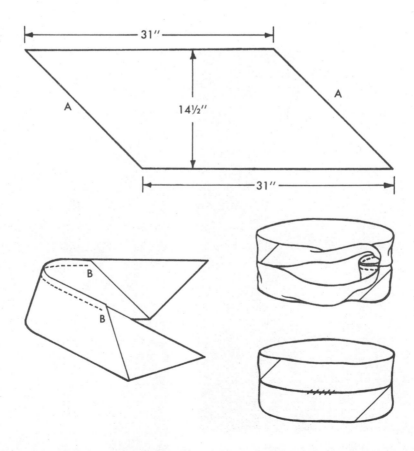

23. The trim on this dress is available by the yard and comes in many colors. It is applied to the dress by hand. Here the same triangular design was used on the front and back of the dress. Note how it was cut and arranged to form the point at the center front. On the belt, a small piece was used to add a finish to the folded ends of the bias belt. The belt fastens at the side with two hooks and eyes.

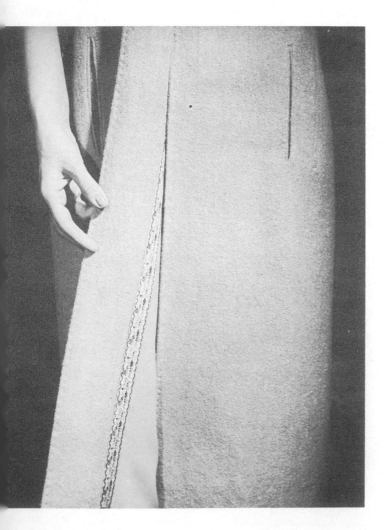

24. A different and expensive touch on a suit or coat is to sew pretty braid at the edge of the facing and lining.

227

25. The most desirable way to put a velvet or velveteen collar on a tailored garment is to complete the garment, and then add the extra collar. This way the velvet collar can be removed when it gets worn-looking.

Baste-stitch the collar the desired width from the edge that the velvet collar will begin. Cut the velvet collar from the top collar pattern but on the bias, cutting away the difference in the size of the top collar and the ⅝-inch seam allowance. Allow a ¼-inch seam on the velvet collar, and staystitch that distance from the edge. Turn under seam allowance on staystitching and press to perfection. Invisibly slipstitch the velvet collar to the top collar, going through stitching on top collar to give it a set-in look.

26. Gingham embroidery is fun and popular. Use it on a blouse, pocket of a shirt, or bodice of a child's dress; or use it as groups of one design to make a border around the skirt. Gingham with ⅛-inch checks is recommended, and the designs are worked in cross stitches and straight stitches. When making cross stitches, top strands should all point in the same direction.

Use three strands of six-strand floss, and starting at A in a white square, follow chart to make design. We suggest that the first motif be embroidered in white, the dog in blue, and the flowers in green, red, blue, and orange, as indicated.

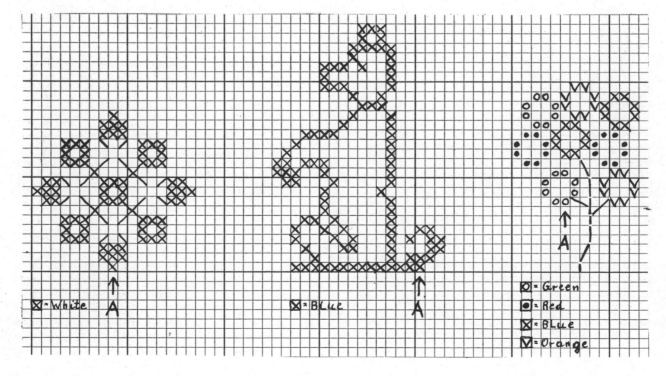

☒ = White   A

☒ = Blue   A

☒ = Green
☒ = Red
☒ = Blue
☒ = Orange

anchor-stitch: stitching about ½ inch long at a given area for reinforcement. Use a short zigzag stitch at machine. If you do not have a zigzag machine, stitch back and forth several times for about ½ inch.

bar tack: reinforcement by stitching at an area of a garment that gets a lot of pull or strain.

baste-stitch: the longest machine stitch, 6 to 8 per inch. With contrasting thread, it is used for basting and for marking location lines for buttonholes, pockets, and center fronts. With matching thread, it is used for control of ease on sleeve caps.

beveling edges: trimming away the seam allowance inside faced edges by using a slanted angle with the scissors to alleviate bulk in heavy fabric.

bias: a diagonal slant in a fabric that does not exactly follow the lengthwise or crosswise grain. True bias makes a 45° angle across the lengthwise and crosswise grain, and has greater stretch and less raveling than any other cut edge.

binding: enclosing both sides of an edge with a strip of fabric.

box plait: finished vertical strip that is sometimes used on the front of shirts. All men's dress shirts have this.

catch-stitch: a hand stitch, taken from right to left at a lower boundary and then at an upper boundary, which forms a triangular design.

clean finishing: turning a raw edge to the inside for ⅛ or ¼ inch on a line of staystitching, and then stitching on the edge. Clean finishing is used to finish the raw edges of facings and hems.

dart: a fold of fabric stitched to give shape to a garment, wide at one end and tapering to a point at the other end.

directional staystitching: a line of regulation machine stitching with matching thread, stitched through a single piece of fabric and placed just outside the seamline unless otherwise stated. It holds the grain threads in position, prevents fabric from stretching, and maintains the pattern line.

directional stitching: stitching seams in the correct direction of the grain to hold the grain threads in position, to prevent the fabric from stretching, and to maintain the pattern line.

drum: lining for a skirt. It is made with a firm fabric and gives support to the skirt. It is not sewn into the seams.

ease: working in extra fabric when stitching to a shorter piece of fabric without having gathers or small tucks.

facing: a piece of fabric cut on grain identical to the garment's, which is used to finish edges, such as necklines and sleeves. Facing may be finished to either the right side or the wrong side of the garment.

favor: ease fabric slightly on front facings above markings for top button and buttonhole.

grain: fabric is woven with the threads interlaced lengthwise and crosswise. The crosswise threads from selvage to selvage form the crosswise grain; the lengthwise threads running parallel with the selvage form the lengthwise grain. When the lengthwise and crosswise threads lie at perfect right angles, the fabric is grain perfect.

hand picking: a running stitch that pricks through the interfacing and penetrates into fabrics with depth. It is used as an edge finish in tailoring. For a zipper the technique is different and is shown on page 41.

interfacing: the fabric that is placed between the inside and the outside of the garment sections, such as in collars and cuffs. Interfacing gives body and better form to the appearance of the finished garment.

lockstitching: knotting the machine thread at the beginning and at the end of a line of stitching by releasing pressure on the presser foot, and stitching several times in the same stitch. Lockstitching eliminates the time and nuisance of tying threads, and the time and motion of using a reverse stitch.

miter: the angle formed when the excess fabric has been removed from a corner by a diagonal seam.

notches: the V-shaped markings on the edge of a pattern to indicate where corresponding pattern pieces are to be joined.

overcast: long, loose stitches over the raw edges of a seam to prevent raveling.

regulation, or permanent, stitch: the permanent stitch that is placed in a garment, 12 to 15 per inch, with matching thread. The number of stitches will vary with the type of fabric; a firm, fine fabric will take a shorter stitch than will a thick, heavy fabric.

running stitch: a hand stitch made by placing the needle in and out of the fabric in an even, straightforward manner.

seam allowance: the portion of a garment allowed for the seam, usually ⅝ inch.

seamline: the exact line where a seam is stitched.

selvage: the two lengthwise finished edges (parallel with the warp threads) on all woven fabric.

shank: the space between button and fabric, on all garments, to give room for the buttonhole side of the garment. It is made with thread unless the button has a metal shank.

short stitch: a shorter machine stitch, 18 to 20 per inch, used for reinforcement on comparatively limited areas, such as points of collars, underarm curves that must be clipped, and tailored buttonholes.

slipstitch: a concealed hand stitch that can be used only on an area with a folded edge. The needle is run along the fold of the hem for about ¼ inch; with a downward movement the garment is pricked; the needle is then brought out at the side of the fold. To continue slipstitching, the needle is placed back into the fold of the hem at about the same place, and the same procedure is repeated.

staystitching plus: manipulating the fabric with the machine stitch to force together the grain threads. It is an easy method for gaining control of ease.

stiletto: a pointed instrument that is used for making eyelet holes, etc.

topstitching: stitching a seam or edge on the right side of a garment with one or more rows of stitching.

tuck: a fold of fabric stitched to give shape to a garment.

underlay: section of fabric attached to drum or underlining of skirt back, forming the underpart of a pleat.

underlining: a second piece of fabric, cut from the same pattern pieces and on the same grain as the garment, staystitched to the outside fabric, and treated as one piece with the outside fabric for further construction. It gives a sculptured and quality look to a garment. When a dress is underlined, interfacings are usually not necessary.

understitching: a row of machine stitching placed close to the edge of any facing, which catches the two trimmed seams to the facing. It keeps the facing to the underside, and sharpens the seam edge.

unit construction: assembling the sections of a garment that make a unit, and completing all stitching and pressing before each unit is joined to another. This process involves less handling, improves organization of work, guarantees better quality-looking clothes, and enables shorter periods of time to be used to advantage.

whipstitch: a hand stitch that is used over a turned edge when sewing together two pieces of fabric.